BEHAVIOURAL ASPECTS OF PARASITE TRANSMISSION

**Supplements to the Journals
of the Linnean Society of London**

Already published
New research in plant anatomy
The full texts of papers read and submitted
for a symposium arranged by the
Society's Plant Anatomy Group and held in
London September 1970. Published as
Supplement 1 to the *Botanical Journal of
the Linnean Society* Vol. **63**, 1970

Early mammals
The full texts of papers read and submitted
for a symposium held in London
in June 1970. Published as Supplement 1
to the *Zoological Journal of the Linnean Society*
Vol. **50**, 1971

Biology and chemistry of the Umbelliferae
Symposium held in September 1970 in conjunction with
the Phytochemical Society and published as
Supplement 1 to the *Botanical Journal of the Linnean
Society* Vol. **64** 1971

Forthcoming titles
**The phylogeny and classification
of the Filicopsida**
Symposium arranged with the British Pteridological
Society held in April 1972

The relationships of fishes
Symposium held in June 1972

Biology of the male gamete
Symposium to be held at St John's College, Cambridge
in September 1973

Published quarterly
Biological Journal of the Linnean Society

Published eight times per year
**Botanical Journal of the Linnean Society
Zoological Journal of the Linnean Society**

EDITED BY **ELIZABETH U. CANNING**

Department of Zoology, Imperial College of
Science and Technology, London

AND **C. A. WRIGHT**

Department of Zoology, British Museum (Natural History),
London

BEHAVIOURAL ASPECTS OF PARASITE TRANSMISSION

Supplement 1 to the Zoological Journal of the
Linnean Society Volume 51 1972

**Published for the Linnean Society
of London by Academic Press**

ACADEMIC PRESS INC. (LONDON) LIMITED
24/28 Oval Road
London NW1

U.S. Edition published by
ACADEMIC PRESS INC.
111 Fifth Avenue
.New York
New York 10003

Library of Congress Catalog Card No. 72 83 797
ISBN 12 158650 2

Printed in Great Britain by
The Whitefriars Press Ltd., London and Tonbridge

Foreword

The Linnean Society being a biological society is in an almost unique position to hold meetings, symposia and discussions upon subjects which are of interest to workers in more than one specialised field. This symposium upon the Behavioural aspects of parasite transmission *was such a meeting, bringing together entomologists, parasitologists, protozoologists and students of animal, including human, behaviour. It was arranged and held in conjunction with the British Society for Parasitology and the British Section of the Society of Protozoologists.*

As this volume of papers read at the symposium shows, it was a successful and profitable meeting for all who attended it. The Society is indeed grateful to the editors, Drs E. U. Canning and C. A. Wright for their part in arranging the symposium and its subsequent publication and also to the Wellcome Trust and the Royal Society for their financial support.

September 1972

DORIS M. KERMACK
Editorial Secretary

v

AUG 4 1981

List of contributors

BETHEL, WILLIAM M. *University of Alberta, Edmonton, Alberta, Canada* (p. 123)

BROOKER, B. E. *The Nuffield Institute of Comparative Medicine, The Zoological Society of London, Regent's Park, London, NW1* (p. 171)

CABLE, RAYMOND M. *Department of Biological Sciences, Purdue University, Lafayette, Indiana 47907, U.S.A.* (p. 1)

COMBES, C. *College Scientifique de Perpignan, Universite de Montpellier, Chemin de Villeneuve, 66-Perpignan, France* (p. 151)

CROLL, NEIL A. *Department of Zoology, and Applied Entomology, Imperial College of Science and Technology, Prince Consort Road, London, SW7* (p. 31)

DUKE, B. O. L. *Helminthiasis Research Unit, Institut de Recherches Medicales, P.O. Box 55, Kumba, Federal Republic of Cameroon* (p. 97)

GATEHOUSE, A. G. *Department of Zoology and Applied Entomology, Imperial College of Science and Technology, Prince Consort Road, London, SW7* (p. 83)

GILLIES, M. T. *School of Biological Sciences, University of Sussex, Falmer, Brighton, Sussex* (p. 69)

HOLMES, JOHN C. *University of Alberta, Edmonton, Alberta, Canada* (p. 123)

LEWIS, C. T. *Department of Zoology and Applied Entomology, Imperial College of Science and Technology, Prince Consort Road, London, SW7* (p. 201)

LLEWELLYN, J. *Department of Zoology and Comparative Physiology, The University, P.O. Box 363, Birmingham 15* (p. 19)

LYONS, KATHLEEN M. *Department of Zoology, King's College, London WC2* (p. 181)

NELSON, G. S. *London School of Hygiene and Tropical Medicine, Keppel Street, London, WC1* (p. 109)

WORMS, M. J. *National Institute for Medical Research, The Ridgeway, London, NW7* (p. 53)

Preface

The study of animal parasites which was for many years largely confined to morphological and systematic investigations has advanced into other fields in a relatively short space of time. Most of this progress has been made possible by technological developments in the fields of biochemical analysis and electron microscopy, permitting a better understanding of the physiological relationships between parasites and their hosts, and extension of morphological observations to a very refined level. Concomitant with these detailed laboratory studies continued interest has been directed towards the life-cycles and transmission patterns of parasites. At one time a life-cycle was considered to be elucidated if the succession of hosts involved was listed in the appropriate sequence and the morphology of the parasite stages in their respective hosts was adequately described. However, attempts to control some parasitic diseases made it clear that more information was needed on the mechanisms by which parasites reach and enter their hosts and this need stimulated an active approach to the ecological background to parasite transmission. A most important factor throughout every phase of these transmission patterns is the behaviour of either the parasites or their hosts and this is an aspect which has received scant attention, perhaps because it has been overshadowed by more dramatic technological advances. In seeking a topic suitable for an interdisciplinary symposium the subject of behaviour recommended itself to the organisers because of the inclusive nature of the interests involved. The papers in this volume review a great deal of the existing knowledge of the behaviour patterns of both parasites and their hosts, they include new contributions on the fine structure of some parasite sense receptors and they provide many examples of the ways in which natural selection operates to maintain the fine balances between parasite and host.

Editors

ELIZABETH U. CANNING
*The British Section of the Society
for Protozoology*

C. A. WRIGHT
*The British Society for Parasitology
and Linnean Society*

Contents

Behaviour of digenetic trematodes

RAYMOND M. CABLE*

Department of Biological Sciences, Purdue University, Lafayette, Indiana, U.S.A.

Trematode behaviour has been studied almost entirely in free miracidia and cercariae, but in one species or another, all stages behave in a manner that seems to favour transmission. Hatching of miracidia and emergence of cercariae may be attuned to habits of the potential hosts which must be reached to survive. Adults of some species retain their eggs and migrate from the vertebrate host to release them. Likewise, germinal sacs may leave the intermediate host with their brood and be eaten by the next host. Cercariae that encyst in the open, often do so attached to the preferred food or prey of the vertebrate host. Excystment of metacercariae is an active process in many species and is triggered by physical and chemical stimuli.

In general, responses of free miracidia and cercariae to physical stimuli bring the larvae into the immediate neighbourhood of potential hosts. There, miracidia may be stimulated to chemokinetic or chemotactic behaviour by substances which are emitted from the host and have been partly identified. Their presence may facilitate host finding or even be essential to penetration but they do not determine host specificity. Chemokinesis or chemotaxis in cercariae before they make contact with a potential host occurs rarely, if at all, but once contact is made, a chemical stimulus triggers penetration in some species at least.

Some unpublished observations on behaviour are reported, including further studies on philophthalmids in which nearly every stage responds to extrinsic stimuli with behaviour that directly or indirectly facilitates transmission from one host species to the next in the life cycle.

CONTENTS

INTRODUCTION

Publications containing no less than three reviews of trematode behaviour have appeared in the non-periodical literature during the past five years, the most recent being in a volume of papers read at a symposium in Toronto last

*Participation in symposium supported in part by travel grant from Purdue Research Foundation and by N.I.H. Grant, No. A109600, U.S. Public Health Service.

1

year. In it, Ulmer (1971) cited literature extensively, reported some original studies and updated his paper until it went to press. Previously, Smyth (1966) and Cheng (1967) gave excellent condensations of some of the more extensive studies on behaviour while Schwabe & Kilejian (1968) reviewed chemical aspects of the subject. Some repetition is desirable, indeed unavoidable, but this paper will dwell more on the behaviour of trematodes that my students and I have met over a period of 40 years, and report some unpublished observations, a few of which were made almost that long ago.

The digenetic trematodes illustrate better than any other group the propriety of a society named after Linnaeus being a sponsor of this symposium on behaviour. Life history studies have demonstrated abundantly the significance of immature stages and their morphology to concepts of trematode taxa from orders to species. Yet the behaviour of those stages can be just as distinctive at all levels, as taxa having fork-tail cercariae perhaps illustrate best. Of features characterizing major taxa, the resting attitude with the tail bent laterally identifies cercariae in the family Cyathocotylidae at a glance. A few years ago, one of that type attracted our attention by its clumsy swimming. It proved to be the long-sought larva needed to demonstrate the complete life history of the only cyathocotylid known to mature in a fish (Stang & Cable, 1966). Finding that cercaria prompted a closer look at the snail host, revealing it to be a scarce species that we had not recognized before as being in the river where we had collected snails for years. If seen, it had been mistaken for an abundant and similar species that was known to harbour another cyathocotylid. Thus did the behaviour of a cercaria contribute to our meagre knowledge of snail systematics.

More than for its contribution to taxonomy, the behaviour of trematodes has been studied for the part it may play in their transmission. Miracidia have received most of the attention but the behaviour of cercariae is more diversified and thereby can be far more obviously adapted to host-finding. Indeed, the behaviour of a parasite could scarcely give a stronger clue to its next host than we have observed in a minute xiphidiocercaria reported here for the first time. Its morphology places it with plagiorchioid larvae, the largest group of cercariae, which as a rule behave in a rather prosaic fashion by swimming almost continuously in no particular direction. Unlike any other in that group to my knowledge, this species swims to the top of the water after emerging from the snail host and attaches to the surface film by the spiny tip of the tail. There the larva remains quiescent except for an occasional wriggle and sometimes detachment for a short swim and then reattachment to the surface film. So obvious were the clues provided by the behaviour of the parasite and by the ecology of its snail host that the first attempt to discover the next host was successful. When placed in dishes of water with cercariae, anopheline mosquito larvae began to feed in the manner distinctive of their kind, pulling in the surface film with cercariae attached. After a short time, feeding stopped and the larvae were plainly in distress. The reason became obvious the next day when each was found to have several metacercariae near the brain. The remainder of the life history still is unknown but delay of encystment until the mosquito larva pupates suggests that the definitive host is infected by eating the adult mosquito and is therefore less likely to be a fish than some other vertebrate.

While recent experimental studies have concentrated on the blood flukes and fasciolids of mammals, our attention has been given to a variety of other trematodes, especially to *Philophthalmus megalurus* after its life cycle was demonstrated and found to have several advantages for experimentation. One is the unusual habitat of the adult in the ocular sac of birds, a site that is not only convenient for the experimenter but also advantageous to the parasite in that its eggs are washed from the bird's eyes when it feeds and are not subject to the risk of being deposited high and dry with the host's faeces. The matter of site selection in *P. megalurus* will be considered further after dealing with other behavioural aspects. The life history is shown in Fig. 1 for reference in further discussion, and to illustrate some of the unusual behaviour facilitating transmission of that species.

REPRODUCTIVE BEHAVIOUR OF ADULTS

Like nearly all trematodes except the schistosomatids, *Philophthalmus megalurus* is hermaphroditic. When alone, the adult readily inseminates itself and produces viable eggs, but self-fertilization rarely occurs when the opportunity of cross-insemination exists as Nollen (1968) showed. Recently, he informed me that he has reared *P. megalurus* through the complete cycle five times, each with single-worm infections as the source of eggs, and has observed no indication of changes in the general vitality at any stage. In other species self-insemination or possibly parthenogenesis (Cable, 1971) results in offspring reported to be partially or fully viable in some species and inviable in others.

The special situation in some dioecious blood flukes in which the female will not develop fully to maturity in the absence of the male has been reviewed by Armstrong (1965). He obtained pairing between abnormal (from X-irradiated cercariae) as well as normal males and females of the same species, heterosexual pairing between species in different genera, and homosexual pairing of males of the same species. His findings indicated that coupling is initiated by thigmotaxis but suggested to him that the male may produce a pheromone essential to the female's attaining sexual maturity and maintaining reproductive function. More recently, Michaels (1969) has postulated a linear arrangement of receptors in both sexes from observations on *in vitro* mating in *Schistosoma mansoni* between the anterior or posterior half of the body in one sex and transections or entire individuals of the other. Anorchid males stimulated females to lay eggs, proving that the presence of sperms or a testicular secretion is not the stimulating agent.

An unusual type of behaviour characterizes certain adult trematodes living where eggs do not have a ready access to the outside. Such a species is *Heronimus mollis*, the American turtle-lung fluke which lays no eggs. Instead, they accumulate in the uterus during hibernation of the host and develop until ready to hatch by early summer when the gravid worm migrates to the outside by way of the trachea and mouth. As water is imbibed, miracidia hatch *in utero* and escape through the genital pore (Crandall, 1960). Even more remarkable is the behaviour of the heterophyid *Acetodextra amiuri*, which matures in the ovaries of catfish and is, to my knowledge, the only trematode known to be cytozoic at any stage; Perkins (1956) found that the young worms invaded the developing eggs of the host. Eggs are retained in the uterus as in the turtle lung

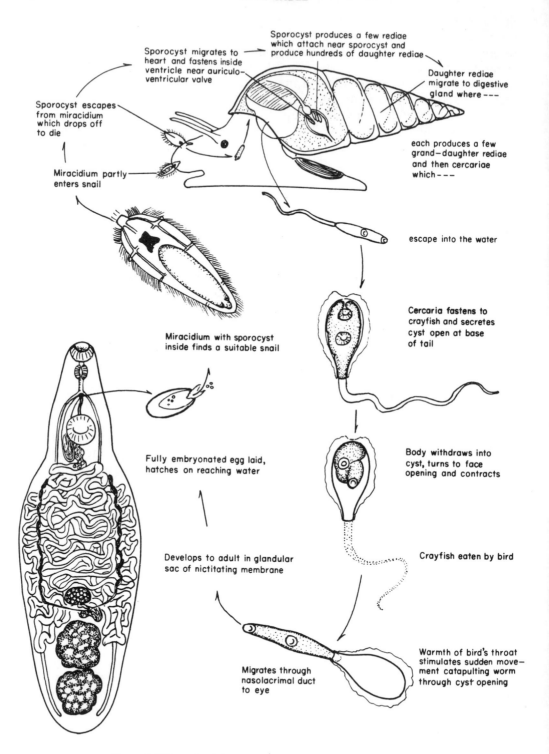

Sporocyst migrates to heart and fastens inside ventricle near auriculo-ventricular valve

Sporocyst produces a few rediae which attach near sporocyst and produce hundreds of daughter rediae

Daughter rediae migrate to digestive gland where ---

Sporocyst escapes from miracidium which drops off to die

each produces a few grand-daughter rediae and then cercariae which ---

Miracidium partly enters snail

escape into the water

Miracidium with sporocyst inside finds a suitable snail

Cercaria fastens to crayfish and secretes cyst open at base of tail

Fully embryonated egg laid, hatches on reaching water

Body withdraws into cyst, turns to face opening and contracts

Crayfish eaten by bird

Develops to adult in glandular sac of nictitating membrane

Migrates through nasolacrimal duct to eye

Warmth of bird's throat stimulates sudden movement catapulting worm through cyst opening

Figure 1. Life cycle of the avian eye fluke, *Philophthalmus megalurus*.

fluke, and the gravid worm is discharged when the host spawns. Less than a minute after reaching the water, the worm ruptures forcibly and ejects a stream of eggs several millimetres into the water. The stimulus is hypotonicity of the water because gravid worms will remain alive and intact in isotonic saline for several days and then release their eggs as described if transferred to water.

BEHAVIOUR OF MIRACIDIA

Hatching

Miracidial behaviour related to transmission begins before hatching from the eggs of many species at least, and probably others including those whose eggs must be eaten by the intermediate host for hatching to occur. That process has been investigated most in *Fasciola hepatica* whose miracidia are activated by light and to a less extent by other stimuli. According to one interpretation, the miracidium releases a hatching enzyme which loosens bonds holding the operculum in place, and permits the miracidium to escape (Rowan, 1956, 1957). Wilson (1968b) offered an alternative explanation suggesting that the activated miracidium alters permeability of a viscous cushion between the larva and the operculum. Imbibing fluid, the cushion swells and exerts pressure on the operculum until it opens and the miracidium is expelled.

Neither of the above interpretations explains satisfactorily the hatching process in blood fluke eggs which are anoperculate and those of other trematodes such as the philophthalmids in which an operculum may be delineated by a faint line but rupture of the shell on hatching may not follow that line. In such trematodes, the egg shell is thin, ruptures on hatching and the older concept attributing hatching to movements of the activated miracidium may be nearer the truth. Such eggs also become fully embryonated before escape from the host and the sudden exposure to water could result in osmotic phenomena concerned with hatching whereas in such species as *Fasciola hepatica*, the eggs require incubation in water for several days to become embryonated and the fully developed miracidium is not exposed to such an osmotic shock just before hatching. It is unlikely, however, that osmotic effects always operate in the same manner, even in all species of the genus *Philophthalmus* with the eggs of some exchanging the ocular sac of birds for the hypotonic environment of fresh water and those of others reaching sea water to hatch. Philophthalmids are unusual also in having eggs that increase in size considerably as they become embryonated but their shells instantly return to near the original size as the miracidium hatches. Such elasticity must exert some pressure helping the larva to rupture the egg shell and escape.

Responses to physical stimuli

After hatching, the behaviour of *Philophthalmus megalurus* miracidia plainly facilitates transmission, first by dispersing the larvae and then by bringing them into proximity with potential snail hosts. In a dish illuminated from the side by a small spotlight, they swim near the surface and directly toward the light on hatching. For some time, we sought to use that behaviour in infecting snails experimentally by exposing them individually to 6-12 miracidia in small dishes arranged in a circle around a lamp in a darkened room. The dishes were turned as necessary to keep the snail on the side towards the light and thus enhance contact with miracidia. The infection rate was erratic and sometimes unbelievably low until it was found that a consistently higher rate could be obtained if just enough water was used to permit the snails to crawl but not leave the bottom of the dish. Freshly collected snails were used because none that can be infected with *P. megalurus* has been reared in satisfactory numbers in captivity. We chose *Pleurocera acuta* because it is hardier, larger and produces more cercariae than other susceptible snails. Its natural environment is the bottom of rivers where it moves to shallow water during the summer but when brought into the laboratory and placed in dishes, it usually climbs as high as possible without leaving the water. Apparently a difference of not over 5 cm in the distance of the snail from the bottom of the water significantly altered the infection rate.

To investigate more precisely the effects of light and water level on infection of *Pleurocera acuta* with *Philophthalmus megalurus,* Mary Schutte and I did some work reported here for the first time. In each of four experiments, freshly collected snails were confined in groups of 20 to 8 cages of which two were placed at each of four depths from just awash at the surface to the bottom of water 37.5 cm deep in an aquarium. The cages were of a size just deep enough to permit the snails to crawl freely and had a combined bottom area close to that of the aquarium. They were placed in it in such a manner that no part of one cage was directly above another, thereby minimizing in the vertical direction any filtering or "decoy" effect such as Chernin (1968) observed when snails were interposed in the path of *Schistosoma mansoni* miracidia moving horizontally in the direction of susceptible "target" snails. After aeration with the cages of snails in place for several hours, several hundred miracidia were added and aeration was continued for 30 s to distribute them evenly in the water. The aquarium was then left undisturbed overnight, either in total darkness or illuminated by a two-tube (20 W each) fluorescent lamp centred 25 cm above the water. The heart of each snail was then removed and examined for sporocysts which migrate to the ventricle and aorta after escape from miracidia which only partly penetrate the snail. The results are given in Table 1. Not included are a few snails which were naturally infected and one cage of 20 placed 17.5 cm from the surface but not examined later.

With illumination from above, most of the sporocysts were in snails confined to cages on the bottom of the aquarium but those kept at the surface yielded almost one-third the total number of sporocysts. The third highest incidence was in cages 5 cm from the bottom while snails kept at mid-depth harboured very few sporocysts. In the dark, results were much the same at mid-depth and 5 cm from the bottom but were strikingly different at other levels. Less than 3% of the snails at the surface were infected with only one sporocyst each, and although the incidence in the bottom cages was about the same as when in the light, the percentage of the total number of sporocysts was much higher because of the greater average number per snail. In similar experiments, Shiff (1969) released miracidia of *Schistosoma haematobium* at a point on the surface of an outdoor pond equidistant from cages of susceptible snails at the surface, mid-level and bottom at a depth of 46 cm with equal numbers in the sun and shade. By far the most snails became infected in the cages at the bottom in the shade.

Table 1. Effects of water level and light on infection of *Pleurocera acuta* with miracidia of *P. megalurus*

Distance from water surface (cm)	No. exposed	No. infected	Snails				Sporocysts Total % of no.	
			No. infected with				at 1 level	at all levels
			1-3	4-6	7-10	10+		
				sporocysts				
0.0								
Light	76	31	24	6	1	0	73	32.0
Dark	78	2	2	0	0	0	2	0.6
17.5								
Light	78	8	8	0	0	0	9	3.9
Dark	57	6	6	0	0	0	6	1.8
32.5								
Light	78	17	17	0	0	0	23	10.9
Dark	73	20	16	3	1	0	42	12.5
37.5								
Light	75	48	37	6	5	0	123	53.9
Dark	77	48	32	10	3	3	288	85.2

From our experience in infecting snails individually with *Philophthalmus megalurus* and watching miracidia during that process, it seemed that they are not inclined to infect the snail for a time after hatching. Behaviour suggesting an increase in the infectivity of miracidia with age has been observed in other species and led Campbell & Todd (1955) to suggest that young miracidia of *Fascioloides magna* do not have a fully developed sensitivity to host stimuli. More recently, Ulmer (1971) has shown that young miracidia of *Megalodiscus temperatus* are not as adept as older ones at host-finding, and suggested that some conflicting observations may be due to behavioural changes with age that have not been considered.

Some further experiments designed to investigate changes in behaviour and infectivity of *Philophthalmus megalurus* miracidia with age have been inconclusive. Observing larvae in vertical migration tubes in the dark except for a small spotlight at the top gave no indication of the rapid descent to the bottom suggested by the low incidence of infection of snails at mid-level of the aquarium in our previous experiments. Perhaps internal reflections in the tube stimulated a behaviour pattern different from that of miracidia in the large aquarium. Using vertical migration tubes separable into three parts for examination, Yasuraoka (1953) introduced young miracidia of *Fasciola hepatica* into the middle portion and recorded their migration into the other two. He attributed to negative geotropism the migration of nearly all of them to the upper level in the dark and their remaining there until the third hour after which the larvae were "accustomed to the effects of gravity" and rather evenly distributed themselves until the 16th hour. Then lowered vitality caused them to drift downward and die. In light from above, positive phototaxis maintained the larvae at near the maximum number in the upper region for the 16 h before their downward movement and death. A similar but less prolonged positive phototaxis by miracidia of the eye fluke could account for the difference that we observed in the infection rate of snails kept at the surface in light and darkness.

Yasuraoka (1953) also investigated rheotaxis and found that miracidia of *F. hepatica* swam directly against a current of up to 1.5 mm/s. Above that velocity, swimming was upstream in a zigzag path, the angle increasing with the current up to a velocity of 3 mm/s above which the larvae were swept away. He suggested that rheotaxis in the respiratory current of the snail host would aid larvae in making contact with it. Webbe (1966) observed a similar response of *Schistosoma mansoni* miracidia to water current which may enhance the scanning capacity of the larvae.

In another paper, Yasuraoka (1954) reported studies on phototactic behaviour of *Fasciola hepatica* miracidia. They swam directly toward light up to a certain intensity above which there was no phototactic response. With monochromatic light at a uniform energy level, the order of effectiveness in eliciting positive phototaxis was green > yellow > blue > red. Sudden increases in intensity caused some miracidia to move away from the light source but not all the way to the opposite end of the observation cell. Yasuraoka also observed that the rate of swimming toward light increased with the temperature between 14° and 32°C.

In a similar study on miracidia of *Schistosoma japonicum*, Takahashi, Mori & Shigeta (1961) observed phototactic responses to differ with age, temperature and light intensity. Phototaxis was positive at low intensities but became negative above a certain intensity which decreased from 2500 lux at 18°C to 10 lux at 34°C. A strong negative geotaxis was affected by light above 5000 lux at temperatures above 20°C, and was overridden at 15°C, by negative phototaxis which caused the miracidia to move to the bottom where the snail host occurred. Chernin & Dunavan (1962) observed the opposite behaviour in *S. mansoni* with a pulmonate snail host which is not a bottom form; negative geotaxis was more effective than negative phototaxis but miracidia located their hosts at considerable distances both horizontally and vertically and did not behave uniformly.

Responses to host stimuli

After responses to physical stimuli bring miracidia into the vicinity of potential hosts, it is generally agreed that their behaviour is influenced by emissions from those hosts but opinion differs as to the type of response and its significance to host-finding. Some see it as chemokinesis and either having no special significance or causing the larvae to swim more rapidly and in a manner which increases the likelihood of chance contact with the host; others regard behaviour leading to that contact as chemotaxis which implies a gradient along which the miracidium moves and not necessarily in a straight line as Wright (1959) has pointed out. The miracidium of *Philophthalmus megalurus* responds to the tissues and juices of the snail host much as described for several other trematodes, and West (1961) observed that when near the snail, it may swim with increased speed "in a tight circle" until it makes contact. However, suitable snail hosts all live in flowing water which is so rapid for some that the accumulation of host emissions to form a concentration gradient of an appreciable length seems unlikely. On the other hand, Ulmer (1971) has shown that the miracidium of *Megalodiscus temperatus* reaches *Helisoma trivolvis*, a pulmonate living predominantly in still water, in a shorter time and often quite directly if, before exposure to infection, the snail is isolated for a time to permit the stimulating agent, presumably in mucus, to diffuse into the water.

When the snail was restrained at the edge of a 9 cm Petri dish for as long as 8 h a miracidium released at the centre of the dish sometimes swam straight to the snail in 15 s as if following a gradient. Ulmer also observed that when placed equidistant between a susceptible snail and another pulmonate refractory to infection, the miracidium showed a decided preference for the normal host species. A preference of miracidia of *Schistosoma mansoni* for the host species over two other pulmonates is implied indirectly by the experiments of Etges & Decker (1963) in testing snail species separately against sham snails. Among other things, Kloetzel (1958) found that when snails were removed from water, something attractive to miracidia was left behind. Plempel (1964) showed that material given off by the snail host would diffuse through a sintered glass filter and attract miracidia of *S. mansoni*. Later, Plempel, Gönnert & Federmann (1966) reported that the attraction was specific so far as *S. mansoni* and *Fasciola hepatica* and their respective snail hosts were concerned.

No specificity or limited specificity of molluscan host products in stimulating behavioural changes in miracidia has also been reported. Wright (1966) tested extracts of snail tissues and products and water in which snails had been left over night. He observed a strong response of *Schistosoma mansoni* miracidia to water conditioned by the host snail and by two other pulmonates in different genera. Extracts of blood and mucus of the host snail had the weakest effect. Chernin (1970) observed in the attraction of miracidia by snail-conditioned water specificity to the extent that water conditioned with oncomelaniid snails attracted *S. japonicum* but not *S. mansoni* or *Fasciola hepatica* miracidia. However, water conditioned with species in two and possibly four genera of pulmonate snails attracted miracidia of all three species. Kawashima, Tada & Miyazaki (1961) observed that miracidia of *Paragonimus ohirai* were attracted to susceptible and refractory snail species in the same genus.

Identification of the miracidium-stimulating substance(s) given off by snails has been approached in different ways. MacInnis (1965) tested various chemicals incorporated singly and in combination in gel pyramids for effects on behaviour of *Schistosoma mansoni* and *Schistosomatium douthitti* miracidia. He found that short-chain fatty acids, some amino acids and a sialic acid attracted miracidia and stimulated them to attach to the gel and attempt to penetrate it. Combinations of those substances were no more effective than they were alone, and in general their effect on miracidia was greater in *S. mansoni* than in *S. douthitti*. Snail tissue rendered ineffective by treatment with water and fat solvents regained the property of attracting miracidia and stimulating penetration after treatment with butyric acid or glutamic acid. More recently, Wilson & Denison (1970b) have used photography to record and quantitate behaviour of *Fasciola hepatica* miracidia and found that C_6 to C_{10} fatty acids increased turning behaviour, the rate of which was affected by chain length, concentration and pH. Increased activity with lower pH was judged to indicate that the acids are more effective in the undissociated state.

Another approach to the problem of identifying stimulating substances is analysis of materials shed by molluscs. Wright (1959) detected specific differences in the mucus from various lymnaeid snails but did not correlate those differences with behaviour. In view of the non-specific response reported for certain parasite-snail combinations and the well known tendency of some miracidia to penetrate unsuitable hosts, differences in the composition of mucus may reflect host specificity after penetration rather than host preference before. Chernin (1970) proposed the term *miraxone* for miracidium-stimulating substances and isolated from snail-conditioned water one that is thermostable, has a molecular weight of less than 500, and retains its activity during prolonged storage. He was unable, however, to demonstrate its presence in any tissue of the snail or in its mucus although that substance has been mentioned repeatedly as the probable source of the stimulating material. Wilson & Denison (1970a) extracted from mucus a "component" with the properties Chernin gave for the miraxone from snail-conditioned water. Previously, Wilson (1968a) found mucus from the snail host of *Fasciola hepatica* to have at least five protein fractions and contain glucose, 16 amino acids and a variety of lipids in concentrations suggesting that several substances may facilitate the host-finding process.

Penetration

Behaviour of miracidia during penetration has been described in species too numerous to cite and several times in *Fasciola hepatica* alone. In that species, Wilson (1970) has applied electron microscopy to the interpretation of the penetration process which is generally considered to be accomplished by the combined action of movements of the miracidium, after attachment by means of the terebratorium, and histolytic secretions of the apical gland and paired cephalic glands. West (1961) believed that the anterior cilia of *Philophthalmus megalurus* miracidia also aid in penetration. Site specificity or preference has been reported. Crandall (1960) observed that miracidia of *Heronimus mollis* showed no tendency to penetrate the delicate surfaces exposed by removing the snail's shell but as usual, the larvae penetrated the epidermis that is normally exposed. He further noted that the miracidium often attempted penetration at several places before it was successful. In some species, however, it seems that such attempts weaken the larva or exhaust essential secretions to the extent that further efforts to penetrate either are not made or are abortive.

BEHAVIOUR OF GERMINAL SACS

In the intermediate host, the trematode cycle includes one to three or more generations of germinal sacs known as sporocysts if they lack a digestive system and rediae if a pharynx and gut are present. The first generation in the molluscan host nearly always is the miracidium transformed into a sporocyst after penetration. The sporocyst contains germinal cells and often embryos which may be already present in the miracidium. In *Heronimus mollis* with probably the simplest life history of any trematode, the germinal cells and embryos carried over from the miracidium develop directly to cercariae and terminate the cycle in the snail. In most other trematodes these embryos become a second sporocyst generation or rediae which in turn may give rise to a third generation of germinal sacs before cercariae are produced.

Cercariae usually escape to continue the cycle in another host but may remain in the mollusc which is eaten by the vertebrate host as occurs in *H. mollis*. Transmission of some trematodes is facilitated by the behaviour of germinal sacs, if not indeed dependent on it. The classic example of such behaviour occurs, of course, in the brachylaemid trematodes which have terrestrial and amphibious pulmonates as intermediate hosts and cercariae that do not become free-living. Instead, they remain in the parent sporocyst and move into its branches which extend into the snail's tentacles and become conspicuously pigmented brood sacs. Although they seem to be devoid of innervation. the sacs pulsate in response to light and temperature changes, luring birds to eat the infected snails or pick off the tentacles containing the brood sacs. Moreover, behaviour of the snail is altered, making it more conspicuously present where and when birds feed (Kagan, 1952).

In a type of behaviour that occurs sporadically in a few families, germinal sacs retain their brood, either as cercariae or transformed to metacercariae, and migrate from the mollusc to be eaten by the next host. Also, germinal sacs may remain in the molluscan host and still play a part in the emergence of cercariae. Many trematodes including *Philophthalmus megalurus* have cercariae that complete their development in germinal sacs. Snails infected with *P. megalurus* begin to shed cercariae in a few minutes after being placed in the dark. West (1961) removed "mature" rediae and found that they released 5-20 times as many cercariae in the dark as in the light, indicating that the light stimulus was not mediated by the host but directly affected the rediae, cercariae or both.

Another type of behaviour concerns infection of the snail host and probably is limited to species in which the miracidium contains on hatching one fully-formed offspring which is a redia in species of *Parorchis* and certain cyclococliids. In *Philophthalmus* species, it differs from a redia only in lacking a pharynx and gut and thus is technically a sporocyst. Instead of penetrating the snail and becoming a sporocyst, the miracidium of *P. megalurus* partly enters the host and the sporocyst escapes to consummate the infection process (Fig. 1). It is not expelled by the miracidium as might be supposed. Instead, West (1961) described the process as one in which the sporocyst becomes activated, pushes the terebratorium aside and escapes, much as the miracidium behaves in hatching from the egg. We have observed activation and escape of the sporocyst *in vitro,* proving that a stimulus from the host is not necessary. It may result from ageing or deterioration of miracidia because sporocysts emerge *in vitro* only after several hours, when swimming has practically ceased. Free in saline, the sporocysts remain exceedingly active and in good condition for several days in the refrigerator.

BEHAVIOUR OF CERCARIAE

Emergence

Some cercariae emerge passively by being expelled from the intermediate host; others take an active part in that process. A frequent adaptation serving their transmission is to emerge at times correlated with the habits of the next host. Yamaguti (1970) recognized that correlation in reviewing the periodicity of emergence and swimming behaviour family by family but mentioned few instances of cercariae in the same family emerging at different times. A well known example is *Schistosomatium douthitti* whose cercariae are shed at night when rodents serving as the final host are most active, and the human schistosomes whose larvae emerge during the day.

To that example may be added two others from our experience. One concerns the Philophthalmidae in which cercariae encyst on food of the vertebrate host. A particular kind is not essential but the surface preferred by *P. megalurus* larvae is the exoskeleton of crayfish which remain largely concealed during the day but come out at dusk. Cercariae emerge at that time. The opposite occurs in another philophthalmid, *Parorchis acanthus,* whose cercariae emerge in the light and are less discriminating as to surfaces for encystment. The second example concerns the undescribed cercaria of *Microphallus opacus,* the metacercaria of which has been reported many times from the digestive gland of crayfish. The cercaria emerges during the early hours of darkness when the intended host is abroad, and swims almost continuously until stimulated mechanically; swimming then stops abruptly. This behaviour favours cercariae being carried passively into the gill chamber of the crayfish where they penetrate the gills and pass with the blood to the digestive gland. Cercariae of another microphallid, *M. nicolli,* have the same swimming behaviour but emerge at any hour and penetrate the blue crab whose activity is correlated more with tidal rhythm than time of day.

Swimming

All known philophthalmid cercariae have a unique type of swimming which Rees (1971) has described in detail. It accomplishes little other than keeping the larvae suspended in the water and is affected largely by jerky dorsoventral movements of the body with the tail more or less trailing. In *P. megalurus,* the "scanning" area is increased by extension of the tail until it is long and slender. First contact with the crayfish may be made by either the body or the tail which fastens its glandular tip to the exoskeleton after which continued lashing of the body results in its contact with the host.

In general, behaviour of cercariae after emerging from the intermediate host consists first of swimming or creeping. As in miracidia, that behaviour serves to bring cercariae and potential hosts together. As a rule, cercariae in each family have characteristic free-living behaviour but, as already intimated, it is so varied from family to family and there are so many exceptions to family patterns that no one yet has presumed to chronicle it fully, and no pretence of doing so is made here. Most observations have been incidental to other studies, but a few investigations have dealt primarily with cercariae. One of the most comprehensive, on *Posthodiplostomum cuticola* by Dönges (1963, 1964), was stressed by Smyth (1966). More recently, Haas (1969) has made a comparable study of *Diplostomum spathaceum.*

The most familiar behaviour of cercariae is their response to light which probably is universal among ocellate cercariae but is by no means limited to them. As marked positive phototaxis as I have ever observed is exhibited by an undescribed echinostome cercaria that lacks eyespots or other body pigment. Nevertheless it swims tail-first toward the light with smooth undulations of a gigantic tail which is pigmented a reddish brown and probably lures small fish to eat the larva as Nasir & Skorza (1968) observed in a similar species. In resting larvae, the tail twitches and pulsates, suggesting the movements of brachylaemid brood sacs which are sensitive to light and similarly pigmented.

Phototactic behaviour of cercariae tends to be positive in cercariae whose next host is a vertebrate, usually a fish, and negative if that host is a bottom-dwelling invertebrate. Sometimes, phototaxis is initially positive and changes later, usually a short time after emerging from the intermediate host. A behaviour typical of resting opisthorchioid cercariae is to swim when shadowed as if by a fish passing between them and the surface of the water. The reverse, i.e., cessation of swimming when shadowed, also has been reported (Haas, 1969).

Penetration

The majority of trematodes by far have cercariae that must penetrate the next host, either from the outside or from an internal passage after being eaten. That host is the definitive host, of course, in blood flukes but more often it is a second intermediate host which is eaten by the definitive host but may be secondarily dropped from the life cycle of some trematodes. Thus in the plagiorchioids, a large group predominantly with three-host cycles, species of *Opisthoglyphe* have cercariae directly infecting frogs which are final hosts. Grabda-Kazubska (1969) found that a species infecting the frog as an adult and one entering it as a tadpole have different behaviour patterns adapted to finding and infecting the two stages.

Unlike miracidia, cercariae show little indication of chemical attraction to potential hosts. In studies with marine opisthorchioid cercariae that penetrate fishes, we found no evidence of chemical attraction exerted beyond the surface of the fish and decided that chance contact of the oral sucker or possibly receptors at the anterior end of the cercaria provided the stimulus to attach and begin penetration. That the stimulus itself was chemical seemed likely from experiments in which cercariae of *Cryptocotyle lingua* showed a strong penetration response to agar gel blocks incorporating expressed fluid from cutaneous tissue and muscle of the cunner which serves as the second intermediate host (Cable & Hunninen, 1942, and unpubl. data). Goodchild (1960) observed that spirochid cercariae "whirled in an agitated fashion" on approaching the head of a turtle. Recently, Millemann & Knapp (1970) reported that cercariae of the salmon-poisoning fluke, although unable to swim, become very active when within 1 mm of the fish and penetrate it at any point. Stirewalt (1970) found that the penetration-stimulating effect on cercariae of *Schistosoma mansoni* was removed from dried skin membranes by extraction with fat solvent, and restored by applying hair or skin lipid to them. Lipid alone had a weak effect and temperature none, but in combination with a temperature differential across membranes, they provided a powerful stimulant to penetration. Using hardened gelatin membranes, Clegg (1969) observed that pure cholesterol was as effective as chicken skin lipid in stimulating penetration by cercariae of a bird schistosome.

Cercariae differ widely in the specificity of the animals that they penetrate. Some will penetrate and encyst in almost anything they can enter, and often do to their detriment. More than one erroneous account of a life history has been due to observing such abortive behaviour and assuming that it is responsible for infective metacercariae occurring naturally in the species penetrated. On the other hand, some cercariae seem to make no effort to penetrate species that are not hosts in which development can occur normally. The seeming contradiction that adult blood flukes are not among the most host-specific of all trematodes as their environment would indicate may be explained by immunological studies (Terry, Clegg & Smithers, 1970).

A great deal has been written on penetration by cercariae, especially schistosome larvae. The process is an active one, of course, but much histochemical and biochemical evidence indicates that it is aided by glandular secretions. Common to all penetrating cercariae are paired cephalic glands, each a cell prolonged to form a duct opening anteriorly. In addition, other secretory structures are characteristic of cercariae in certain families: the virgula "organ" in the Lecithodendriidae; caudal glands in the Opecoelidae and Philophthalmidae; the post-acetabular mucin gland in the Troglotrematidae. In several families, the cephalic glands all appear to have the same composition and function; in others the pairs differ in microscopic structure, staining reactions, and probably function.

Mucin secreted by some of the glands is believed to have protective, adhesive and lubricating properties, and is reported in some cercariae to form, on the outside of the second intermediate host, temporary "cysts" from which the cercarial body penetrates the surface (Kruidenier, 1951; Hall & Groves, 1963). The adhesive function of caudal glands in philophthalmid larvae is obvious from casual observation but Porter & Hall (1970) suggested still another function for them in opecoelid cercariae: supplying metal ions as co-enzymes of apoenzymes in secretions of the cephalic glands. Several enzymes have been reported for cercariae but opinion differs as to their identity. Their source has not been established but must be in part at least the cephalic glands whose secretions have a softening or lytic effect during penetration.

BEHAVIOUR OF METACERCARIAE

After metacercariae reach suitable final hosts, the infection process begins with excystment except in trematodes whose metacercariae remain free in the intermediate host. Because excystment usually occurs in the duodenum and is accelerated by the host's digestive enzymes, these substances have commonly been used to excyst metacercariae *in vitro*, especially those in fishes. In *Fasciola hepatica,* however, Dixon (1966) reported that pepsin and trypsin are neither able to accomplish excystment unaided, nor indispensable to that process. Instead, metacercariae are activated by the combination of pCO_2, reducing conditions and temperature normally encountered in the host's stomach; they are then stimulated to emerge by bile and conditions met in the duodenum. Although the requirement for none of those factors was absolute, Howell (1968) found bile essential to excystment in *Echinoparyphium serratum.* Dixon had observed characteristic behavioural changes with activation and during emergence from the cyst and suggested that enzymes from the worm itself may weaken or dissolve the pre-formed ventral plug at the site of emergence.

In unpublished studies, we have found that metacercariae of *Microphallus opacus* would excyst in less than 10 min when placed in trypsin after treatment with pepsin, both at 40°C and their optimum pH values. Excystment occurred spontaneously after several hours in Earle's saline alone. Adding 10% fresh ox bile did not shorten the time appreciably although bile has a pronounced activating effect on a variety of helminths.

Fine structure of the metacercarial cyst differs greatly between such species as *Fasciola hepatica,* in which the cyst forms quickly and completely from materials secreted by the worm, and in trematodes whose cercariae penetrate second intermediate hosts and have a cyst that is thin at first but gradually thickens with age. With those differences, there is little basis for expecting the excystment process to be the same, or to be triggered by the same stimulus in the two cyst types. In current studies on fine structure, Mary Schutte has found the cysts of *Philophthalmus megalurus* and *Parorchis acanthus,* species in the same family, to be much less alike than are cysts of *P. acanthus* and *F. hepatica* which are in different families. A feature common to the last two species and to trematodes in several other families is a type of cystogenous gland containing rod-like bodies or bâtonnets. In electron micrographs of *P. acanthus,* they appear as keratin scrolls which unroll and are applied one after the other to form the laminated inner layer, precisely as Dixon & Mercer (1964) described in *F. hepatica.* As Fig. 1 shows, the form of the cyst and encystment behaviour is unique in species of *Philophthalmus* although their cercariae are indistinguishable from larvae of *Parorchis* species except that none of the cystogenous glands are of the bâtonnet-containing type. Hence we did not expect the cyst of *P. megalurus* to have a laminated inner membrane. That deficiency and an opening in the cyst through which the worm escapes, usually with the first lunge in response to the warmth of the bird's throat, are obviously specializations associated with the unusual habitat of the adult in the ocular sac of birds.

After metacercariae become free in the definitive host, the infection is not consummated until the worm reaches the final site at which the rest of its life is to be spent. It may be near that site when it excysts but usually is not and must migrate to it. The route is often devious and some do go astray; the swim bladder of physoclistic catfishes was regarded as the normal location of *Acetodextra amiuri* adults long before the ovaries were found to be the real site. Because migration and site-finding in the definitive host have been well summarized by both Smyth (1966) and Ulmer (1971), some further remarks on that subject from the standpoint of our current research will suffice.

In connection with attempts to cultivate *Philophthalmus megalurus in vitro,* David Danley has determined that although adults live within a pocket on the inner surface at the base of the nictitating membrane, the actual habitat of the worms for a time after reaching the eye is the acini of the Harderian gland whose duct opens into that pocket. He has removed the glands from chick embryos and maintained them for a time in culture media. When excysted metacercariae were added, they quickly found the gland duct and could be followed moving up it until they entered the acini and began to feed. In contrast to that habitat, other philophthalmid trematodes are intestinal parasites, with species of *Parorchis* living in the bursa Fabricius and cloaca of the bird. Although anatomically remote from one another and seemingly with little in common, these sites may have a functional relationship bearing on their "selection" by species of *Philophthalmus* and *Parorchis.* Mueller, Sato & Glick (1971) have reported that bursectomy interferes with the normal development of the Harderian gland which seems to be dependent on the bursa for plasma cells. The potential significance of that relationship to site "selection" and our cultivation studies is currently under investigation.

REFERENCES

ARMSTRONG, J. C., 1965. Mating behavior and development of schistosomes in the mouse. *J. Parasit.*, *51:* 605-616.

CABLE, R. M., 1971. Parthenogenesis in parasitic helminths. *Am. Zool.*, *11:* 267-272.

CABLE, R. M. & HUNNINEN, A. V., 1942. Studies on the life history of *Siphodera vinaledwardsii* (Linton) (Trematoda : Cryptogonimidae). *J. Parasit.*, *28:* 407-422.

CAMPBELL, W. C. & TODD, A. C., 1955. Behavior of the miracidia of *Fascioloides magna* (Bassi, 1875) Ward, 1917, in the presence of the snail host. *Trans. Am. microsc. Soc.*, *74:* 342-346.

CHENG, T. C., 1967. Marine molluscs as hosts for symbioses. *Advances in marine biology*, *5:* xiii + 424 pp. New York: Academic Press.

CHERNIN, E., 1968. Interference with the capacity of *Schistosoma mansoni* miracidia to infect the molluscan host. *J. Parasit.*, *54:* 509-516.

CHERNIN, E., 1970. Behavioral responses of miracidia of *Schistosoma mansoni* and other trematodes to substances emitted by snails. *J. Parasit.*, *56:* 287-296.

CHERNIN, E. & DUNAVAN, C. A., 1962. The influence of host-parasite dispersion upon the capacity of *Schistosoma mansoni* miracidia to infect *Australorbis glabratus. Am. J. trop. Med. Hyg.*, *11:* 455-471.

CLEGG, J. A., 1969. Skin penetration by cercariae of the bird schistosome, *Austrobilharzia terrigalensis*; the stimulating effect of cholesterol. *Parasitology*, *59:* 2P-3P.

CRANDALL, R. B., 1960. The life history and affinities of the turtle lung fluke, *Heronimus chelydrae* MacCallum, 1902. *J. Parasit.*, *46:* 289-307.

DIXON, K. E., 1966. The physiology of excystment of the metacercaria of *Fasciola hepatica* L. *Parasitology,* *56:* 431-456.

DIXON, K. E. & MERCER, E. H., 1964. The fine structure of the cyst wall of the metacercaria of *Fasciola hepatica* L. *Q. Jl microsc. Sci.*, *105:* 385-389.

DÖNGES, J., 1963. Reizphysiologische Untersuchungen an der Cercarie von *Posthodiplostomum cuticola* (v. Nordmann, 1832) Dubois, 1936, der Erreger des Diplostomatiden-Melanoms der Fische. *Verh. dt. zool. Ges.*, 216-223.

DÖNGES, J., 1964. Der Lebenzyklus von *Posthodiplostomum cuticola* (von Nordmann, 1832) Dubois, 1936 (Trematoda : Diplostomatidae). *Z. Parasitenk.*, *24:* 169-248.

ETGES, F. J. & DECKER, C. L., 1963. Chemosensitivity of the miracidium of *Schistosoma mansoni* to *Australorbis glabratus* and other snails. *J. Parasit.*, *49:* 114-116.

GRABDA-KAZUBSKA, B., 1969. Studies on abbreviation of the life cycle in *Opisthoglyphe ranae* (Fröhlich, 1791) and *O. rastellus* (Olsson, 1876) (Trematoda : Plagiorchiidae). *Acta parasit. pol.*, *16:* 249-269.

HAAS, W., 1969. Reizphysiologische Untersuchungen an Cercarien von *Diplostomum spathaceum. Z. vergl. Physiol.*, *64:* 254-287.

HALL, J. E. & GROVES, A. E., 1963. Virgulate xiphidiocercariae from *Nitocris dilatatus* Conrad. *J. Parasit.*, *49:* 249-263.

HOWELL, M. J., 1968. Excystment and *in vitro* cultivation of *Echinoparyphium serratum. Parasitology*, *58:* 583-597.

KAGAN, I. G., 1952. Further contributions to the life history of *Neoleucochloridium problematicum* (Magath, 1920) new comb. (Trematoda : Brachylaematidae). *Trans Am. microsc. Soc.*, *71:* 20-44.

KAWASHIMA, K., TADA, I. & MIYAZAKI, I., 1961. Host preference of miracidia of *Paragonimus ohirai* Miyazaki, 1939, among three species of snails of the genus *Assiminea. Kyushu J. med. Sci.*, *12:* 99-106.

KLOETZEL, K., 1958. Observações sobre o tropismo de miracidio de *Schistosoma mansoni* pelo molusco *Australorbis glabratus. Revta bras. Biol.*, *18:* 223-232.

KRUIDENIER, F. J., 1951. The formation and function of mucoids in virgulate cercariae, including a study of the virgula organ. *Am. Midl. Nat.*, *46:* 660-683.

MacINNIS, A. J., 1965. Responses of *Schistosoma mansoni* miracidia to chemical attractants. *J. Parasit.*, *51:* 731-746.

MICHAELS, R. M., 1969. Mating of *Schistosoma mansoni* in vitro. *Expl Parasit.*, *25:* 58-71.

MILLEMANN, R. E. & KNAPP, S. E., 1970. Biology of *Nanophyetus salmincola* and "salmon poisoning" disease. *Adv. Parasitol.*, *8:* 1-41.

MUELLER, A. P., SATO, K. & GLICK, B., 1971. The chicken lacrimal gland, gland of Harder, cecal tonsil and accessory spleens as sources of antibody-producing cells. *Cell. Immunol.*, *2:* 140-152.

NASIR, P. & SKORZA, J. V., 1968. Studies on freshwater larval trematodes. XVIII. The life cycle of *Stephanoprora denticulata* (Rudolphi, 1802) Odhner 1910 (Trematoda : Digenea : Echinostomatidae). *Z. Parasitenk.*, *30:* 134-138.

NOLLEN, P. M., 1968. Autoradiographic studies on reproduction in *Philophthalmus megalurus* (Cort, 1914) (Trematoda). *J. Parasit., 54:* 43-48.

PERKINS, K. W., 1956. Studies on the morphology and biology of *Acetodextra amiuri* (Stafford) (Trematoda : Heterophyidae). *Am. Midl. Nat., 55:* 139-161.

PLEMPEL, M., 1964. Chemotaktische Anlockung der Miracidien von *Schistosoma mansoni* durch *Australorbis glabratus. Z. Naturf., 196:* 268-269.

PLEMPEL, M., GÖNNERT, R. & FEDERMANN, M. 1966. Versuche zur Wirtsfindung von Miracidien. *1st Int. Congr. Parasit., Rome,* 32.

PORTER, C. W. & HALL, J. E., 1970. Histochemistry of a cotylocercous cercaria. I. Glandular complex in *Plagioporus lepomis. Expl Parasit., 27:* 378-387.

REES, G., 1971. Locomotion of the cercaria of *Parorchis acanthus* Nicoll and the ultrastructure of the tail. *Parasitology, 62:* 489-503.

ROWAN, W. B., 1956. The mode of hatching of the egg of *Fasciola hepatica. Expl Parasit., 5:* 118-137.

ROWAN, W. B., 1957. The mode of hatching in the egg of *Fasciola hepatica.* II. Colloidal nature of the viscous cushion. *Expl Parasit., 6:* 131-142.

SMYTH, J. D., 1966. *The physiology of trematodes,* xv + 256 pp. London: Oliver & Boyd.

SCHWABE, C. W. & KILEJIAN, A., 1968. Chemical aspects of the ecology of platyhelminths. *Chem. Zool., 2:* 467-549. New York: Academic Press.

SHIFF, C. J., 1969. Influence of light and depth on location of *Bulinus (Physopsis) globosus* by miracidia of *Schistosoma haematobium. J. Parasit., 55:* 108-110.

STANG, J. C. & CABLE, R. M., 1966. The life history of *Holostephanus ictaluri* Vernberg, 1952 (Trematoda : Digenea) and immature stages of other North American freshwater cyathocotylids. *Am. Midl. Nat., 75:* 405-415.

STIREWALT, M. A., 1970. Analysis of cercarial penetration stimuli. *J. Parasit., 56* (4, Sec. II): 330-331.

TAKAHASHI, T., MORI, K. & SHIGETA, Y., 1961. Phototactic, thermotactic and geotactic responses of miracidia of *Schistosoma japonicum. Jap. J. Parasit., 10:* 686-691.

TERRY, R. J., CLEGG, J. A. & SMITHERS, S. R., 1970. Host antigens associated with Schistosomes. *J. Parasit., 56* (4, Sec. II): 340-341.

ULMER, M. J., 1971. Site-finding behavior in helminths in intermediate and definite hosts. In A. M. Fallis (Ed.), *Ecology and physiology of parasites,* x + 258 pp. University of Toronto Press.

WEBBE, G., 1966. The effect of water velocities on the infection of *Biomphalaria sudanica tanganyicensis* exposed to different numbers of *Schistosoma mansoni* miracidia. *Ann. trop. Med. Parasit., 60:* 85-89.

WEST, A. F., 1961. Studies on the biology of *Philophthalmus gralli* Mathis and Leger, 1910 (Trematoda : Digenea). *Am. Midl. Nat., 66:* 363-383.

WILSON, R. A., 1968a. An investigation into mucus produced by *Lymnaea truncatula,* the snail host of *Fasciola hepatica. Comp. Biochem. Physiol., 24:* 629-633.

WILSON, R. A., 1968b. The hatching mechanism of the egg of *Fasciola hepatica* L. *Parasitology, 58:* 79-89.

WILSON, R. A., 1970. Penetration of the miracidium of *Fasciola hepatica* into its intermediate host. *J. Parasit., 56* (4, Sec. II): 484.

WILSON, R. A. & DENISON, J., 1970a. Studies on the activity of the miracidium of the common liver fluke, *Fasciola hepatica. Comp. Biochem. Physiol., 32:* 301-313.

WILSON, R. A. & DENISON, J., 1970b. Short chain fatty acids as stimulants of turning activity by the miracidium of *Fasciola hepatica. Comp. Biochem. Physiol., 32:* 511-517.

WRIGHT, C. A., 1959. Host location by trematode miracidia. *Ann. trop. Med. Parasit., 53:* 288-292.

WRIGHT, C. A., 1966. Miracidial responses to molluscan stimuli. *Int. Congr. Parasit. (1st), Rome:* 1058-1059.

YAMAGUTI, S., 1970. On the periodicity of natural emergence of cercariae. *H. D. Srivastava Comm. Vol.,* 485-492.

YASURAOKA, K., 1953. Ecology of the miracidium. I. On the perpendicular distribution and rheotaxis of the miracidium of *Fasciola hepatica* in water. *Jap. J. med. Sci. Tech., 6:* 1-10.

YASURAOKA, K., 1954. Ecology of the miracidium. II. On the behavior to light of the miracidium of *Fasciola hepatica. Jap. J. med. Sci. Tech., 7:* 181-192.

DISCUSSION

C. A. Wright

Is there any evidence for responses to thermal stimuli in cercariae other than those of the schistosomes?

R. M. Cable

None to my knowledge other than an increased rate of development in the molluscan host at higher temperatures. Except in the schistosomes, the response of free-swimming cercariae to thermal stimuli would seem to have little if any bearing on transmission because of the rarity of homoiothermal second intermediate hosts.

J. C. Holmes

Did you notice any change in the behaviour of the anopheline larvae which had plagiorchioid metacercariae next to the brain?

R. M. Cable

Only the temporary one seemingly associated with penetration by cercariae and their localization in the head capsule. After feeding stopped, the larvae writhed for some time, turning the head and retracting it strongly against the thorax.

H. D. Chapman

Is it possible that future work could be directed to critical quantitative studies of trematode behaviour?

R. M. Cable

Yes, but technical difficulties would restrict electrophysiological studies. Other aspects of trematode behaviour would seem to lend themselves well to quantitative studies.

N. A. Croll

There is no evidence that taxes or kineses should junction singly and one of the problems is to separate categories of response during analyses.

Behaviour of monogeneans

J. LLEWELLYN

Department of Zoology and Comparative Physiology,
The University, Birmingham

The fish hosts of monogeneans may swim up to a few hundred times as fast as the invasive larvae of these parasites, and it is likely that invasion takes places when the fishes are for some reason stationary in the vicinity of those sites on the sea or river bottom where parasite eggs have been laid previously. Phototaxis, chemotaxis and rheotaxis are known to play a part in the behaviour of various monogenean larvae, and invasion may be by direct contact with the body of the host in skin parasites, or via the skin and operculum or via the mouth in gill parasites. Some monogeneans occupy, as juveniles, different microhabitats from those occupied as adults.

CONTENTS

INTRODUCTION

Most monogeneans are parasites of the skin and gills of fishes, are species-specific to their hosts, have a single-host life-cycle, and apart from some exceptions to be discussed later, are transmitted to new hosts by means of a short-lived free-swimming ciliated oncomiracidium which has undergone embryonation over a period varying from days to months in open water.

Oncomiracidia have been found to swim at speeds of up to 5 mm/s (see p. 24), while fishes of 10-25 cm length can cruise at about 300-1500 mm/s respectively (i.e. 3-6 lengths/s, see Bainbridge, 1960). Thus the host fishes may, in their normal swimming, travel from about 60 to 300 times as fast as the fastest speed that appears yet to have been recorded for monogenean free-swimming larvae.

Even if the fish swam on a straight course and the larva possessed excellent long-distance target-locating receptors, this speed differential would make it extremely unlikely that successful invasion could normally be the result of the parasite intercepting its specific host while the fish was cruising. The number of occasions on which the larva would be (a) in a position to intercept a particular fish and (b) able to swim fast enough to "home" on to it would scarcely improve the chances of collision above the random level. Thus it appears reasonable to expect that the reproductive habits and behaviour of the parasites will have become adapted to some feature of the host's habits and behaviour so that invasion is attempted when the host fishes are vulnerable, e.g. when they are resting, shoaling, congregating for spawning, or concentrated in localized feeding grounds.

The information available to test the above hypothesis is fragmentary; very little indeed is known about the invasion of hosts by monogeneans, and the present brief review draws heavily on a few comparatively recent studies, namely those of Bovet (1967) on *Diplozoon paradoxum* invading the bream *Abramis brama* and the roach *Rutilus rutilus*; Kearn (1967a) on *Entobdella soleae* invading the sole *Solea solea*; Combes (1968) on species of *Polystoma* invading tadpoles of frogs belonging to the genera *Rana*, *Pelobates* and *Hyla*; and Paling (1969) on *Discocotyle sagittata* invading the trout *Salmo trutta*. Considerable use has also been made of Bychowsky's (1957) extensive account of the general biology of monogeneans.

CONDITIONS PROMOTING THE OCCURRENCE OF PARASITE LARVAE
AND HOSTS IN THE SAME PLACE AT THE SAME TIME

Relation of time of egg production of parasite to habits of host

A common feature of the mode of life of many fishes is to congregate inshore during a restricted part of the year for spawning and then to disperse afterwards into open waters, leaving the developing young in inshore waters. Thus there is often a segregation, resulting in little or no contact, between the young and old fishes of a particular species. Associated with this segregation, some parasites have become adapted to living on the young fishes concentrated in the nursery grounds, while others live exclusively on the older fishes dispersed in the open sea, infection of the latter taking place during the spawning assembly.

Protancyrocephalus strelkovi parasitizes *Limanda aspera* in the Far East, lives less than one year, and is restricted to those fishes which inhabit the littoral zone, i.e. to those fishes under three years old; older fishes, though they migrate towards the shore for spawning in summer, do not go in far enough to expose themselves to infection (Bychowsky, 1957). In *Gastrocotyle trachuri,* a gill parasite also having an annual life-cycle, the reproductive behaviour is such that the oncomiracidia invade *Trachurus trachurus* when the young fishes (in their first two years of life) remain continuously in the inshore nursery grounds. Over 95% of these young fishes are infected, each bearing about a dozen parasites. When sexually mature the fishes migrate to the open sea, returning to the nursery grounds only to spawn. Thus these older fishes spend only a restricted period in inshore waters, and the chances of them becoming infected by *Gastrocotyle* are correspondingly small, resulting in only about 50% of them being infected, and with each fish bearing no more than two or three parasites. A further feature limiting the period of infection of *Trachurus* of all ages is that the larvae of *Gastrocotyle* do not invade the hosts in summertime when the fishes are mainly planktonic feeders, but only when the fishes return to bottom-feeding in the autumn (Llewellyn, 1962).

In contrast to the habit of parasitizing younger fishes practised by *Protancyrocephalus* and *Gastrocotyle,* some monogeneans parasitize older fishes. *Mazocraes alosae* lives on species of *Alosa,* all of which make an inshore spawning migration which may extend even into fresh water. There is a marked seasonal rhythm in the reproductive activity of *Mazocraes*; eggs are not produced until the middle of May, but with embryonic development taking only about ten days, larvae invade new hosts by the end of May. Thus the period of infection is very restricted and is synchronized with the middle of the inshore spawning assembly of *Alosa* (April to June) and, since the young fishes do not associate with the older fishes, the infection is confined to the older spawning fishes of three years old or more (Bychowsky, 1957). A somewhat parallel situation occurs in *Discocotyle sagittata* on trout in Lake Windermere. Here the monogenean may live to an age of three or four years, and parasitizes older fishes in the lake itself and not the younger trout in the nursery streams (Paling, 1965).

Fate of the eggs

According to Bychowsky (1957), monogeneans may be divided into two groups characterized by the fate of their eggs: those in which the eggs remain secured, either mechanically or by adhesion, to the body of the host; and those in which the eggs become free and sediment to the bottom.

In the first group Bychowsky included *Nitzschia sturionis* where the eggs stick by means of the expanded ends of their stalks to the lining of the buccal cavity of sturgeons, and *Mazocraes, Octostoma* and (quoting Zeller) *Diplozoon paradoxum,* where the eggs are devoid of stalks or filaments but nevertheless were said to remain in large numbers on the gills of their respective hosts. However, Bovet (1967), after a very extensive study of many aspects of the biology of *Diplozoon,* said that when eggs are found in the gill chamber of the host, their presence is atypical and accidental. With regard to *Mazocraes* and *Octostoma,* Bychowsky drew attention to the similarity in the manner of egg deposition and to the close taxonomic affinity of these monogeneans; however, during the course of collecting hundreds of specimens of *Kuhnia* (=*Octostoma*) *scombri* from *Scomber scombrus* at Plymouth, I have never found eggs on the gills. Moreover, the only monogenean eggs that have been found free in the plankton at Plymouth have been those of *Kuhnia scombri.* It could be that at the times when I collected *Kuhnia* from the mackerel, egg-laying was not in process, and that the subsequent production of eggs when the parasites were placed in dishes was triggered by some unknown factor in the laboratory conditions. However, in another monogenean species *Gastrocotyle trachuri,* I have kept some of the host specimens *Trachurus trachurus* in a small aquarium and then collected parasite eggs from the sediment on the bottom of the tank, indicating that the *Gastrocotyle* specimens on the *Trachurus* gills were in fact producing eggs, yet on subsequent examination of the gills no adherent eggs were found. On very rare occasions, notably one on which several dozen eggs of *Anthocotyle merluccii* were found on the gills of a dead specimen of *Merluccius merluccius,* I have found eggs to accumulate on the gills when active parasites happen to be relatively free of host mucus and when the fish happens to be lying in such a way that the eggs do not slide off; undoubtedly under natural conditions such eggs would have been washed away by the gill ventilating currents of the fish which are known to be powerful enough to have promoted the development of elaborate attachment organs and of asymmetry of the body in many monogeneans.

Thus the attachment of the eggs of *Nitzschia* to the mouth cavity of sturgeons as observed by Bychowsky is perhaps the only authentic example of monogenean eggs being attached to the host of the parent parasite. In *Entobdella soleae* filaments with sticky droplets are present, but these have been shown by Kearn (1963) to attach the eggs to sand grains and not to the body of the host. In some other monogeneans the egg stalks (i.e. the proximal extensions from the egg capsule) have been thought to be associated with the habit of retaining the eggs in the uterus of the parent rather than with the retention of the eggs on the gills of the host (Llewellyn & Tully, 1969).

In those parasite-host systems where perhaps the parasite eggs do remain on the host, it is difficult to see the advantage of the habit to the parasite. If the egg remained on the host until full-term development, then the oncomiracicium would obviously be released in a position favouring re-infection of the same host specimen, though the existence of an initial non-infective stage in the life of the free-swimming larva (see p. 25) would substantially decrease this possibility. In parasites of shoaling fishes, the attached parasite eggs would be carried to sites where there would always be other potential hosts, but unless the whole shoal came to rest from time to time, and hatching took place while the shoal was in one of these resting periods, the relatively slow swimming speed of the larvae (see p. 24) would greatly reduce the chances of successful invasion.

Thus it seems very probable that in the vast majority of monogeneans the eggs are released into open water, and such eggs are dense enough to sediment to the bottom, even in the sea, though the effect of currents on the buoyancy of the eggs has not been investigated. It would appear then, that most monogeneans infect their hosts when the hosts become relatively stationary near the bottom in the restricted localities where eggs happen to have been laid.

Incubation period of the eggs

Usually the period required for the embryonic development of monogeneans varies according to the species and the temperature, but it has been said (Bychowsky, 1957) that this is not always so: sometimes it varies even at a constant temperature, and conversely, sometimes it is constant even when the temperature varies. When development is temperature-dependent, the difference in the incubation time may vary very substantially: e.g. it is about three days in *Dactylogyrus* at 22°-23°C (Prost, 1963), and 16-21 weeks in *Dictyocotyle coeliaca* at 10°C (Kearn, 1970). The special relationship, if any, of this kind of variation to the infection of the particular hosts of these two species is not known, but in *Polystoma integerrimum* the length of the incubation period is one of the factors in the complex situation which results in the simultaneous occurrence of the larvae of the parasite and the susceptible tadpole stage of the host.

In *Dactylogyrus anchoratus* the temperature-dependent embryonation of the eggs may result in some eggs not completing their development before the end of the summer, but such eggs may persist through the winter and, provided they do not actually freeze, hatch in the following spring (Bychowsky, 1957; Prost, 1963).

Hatching of the eggs

When in laboratory culture, most monogenean eggs appear to hatch spontaneously when embryonic development has been completed, but others, e.g. *Leptocotyle minor* (see Kearn, 1965) and *Rajonchocotyle emarginata* (see Wiskin, 1970), do not. It should be appreciated, however, that the examination of eggs in the laboratory usually involves illuminating them, and this itself might provide a hatching stimulus.

Clearly the requirements of a stimulus from the host to release the short-lived larva from its egg capsule would have survival value for the parasite, and there is in fact some evidence for the occurrence of hatching factors. In *Diplozoon paradoxum*, though light is apparently not essential for hatching, the embryos are nevertheless activated by light, and their activity appears to bring about the actual opening of the operculum (Bovet, 1967); and in some species of *Axine*, hatching may be delayed for a long period by placing the eggs in the dark, following which emergence may be stimulated by transfer to the light (Bychowsky, 1957). Presumably this response of embryos to light is linked in some way with the behaviour of the hosts: perhaps these are nocturnal feeders and become more accessible to invading oncomiracidia on coming to rest at daybreak. Another factor which may influence hatching is the turbulence of the water, as observed in *Diplozoon paradoxum* by Bovet (1967), who thought that in nature a triggering of hatching might be brought about by the disturbance of the water due to the host fish bream coming to feed at the bottom of the lake. The existence of a chemical hatching factor derived from the host has been demonstrated in *Squalonchocotyle torpedinis* by Euzet & Raibaut (1960), but Wiskin (1970), though failing to get fully-embryonated eggs of the related *Rajonchocotyle emarginata* to hatch, failed also to find any evidence for a chemical hatching factor in the gills or mucus of the host. It seems that chemical and turbulent factors may be related to the *presence* of a host in the vicinity of a monogenean egg, whereas a light hatching stimulus may be more broadly related to the general *habits* of the host.

BEHAVIOUR OF THE LARVAE

Behaviour of the free-swimming oncomiracidia

Manner of swimming

Once free from the egg capsule, oncomiracidia swim in a forward direction by means of their epidermal cilia; unlike ciliated protozoans, monogenean larvae cannot swim backwards. As it goes forwards, the oncomiracidium usually takes a spiral path (*Entobdella soleae, Diplozoon paradoxum,* and *Discocotyle sagittata* as described by Kearn, 1967a, Bovet, 1967, and Paling, 1969 respectively), but *Discocotyle* may sometimes proceed in a straight line. The spirals performed by *Entobdella* may be so tight that the larva circles round a fixed point, and according to Owen (1970), *Discocotyle* may perform whirling movements and somersaults. While mean speeds have been calculated (5 mm/s at 20°C for *Entobdella,* 3 mm/s for *Diplozoon,* 0.4 mm/s for *Discocotyle*), all the investigators working on these parasites have noted the frequent changes in the rate of swimming, and have observed that from time to time the oncomiracidia may rest or "crawl" or "glide" over the bottom.

Longevity and limit of swimming distance

The total period of free swimming has been found to be 20-30 h at 7°C and 9-14 h at 17°C in *Entobdella,* up to 10 h in *Diplozoon,* and 4-6 h in *Discocotyle.* In the very unlikely event of these larvae keeping on a straight course, and if it is assumed that they are in still water, the maximum distances they could reach from the site of hatching would be very approximately 300 m for *Entobdella,* 100 m for *Diplozoon,* and 10 m for *Discocotyle.* In life it is most unlikely that the larvae would swim these distances from the eggs, and Bychowsky (1957) observed that in *Polystoma integerrimum,* tadpoles which had drifted only 2 to 3 m from where they had hatched remained uninfected, whereas those tadpoles which had not drifted from their place of hatching (i.e. had remained at the site where both frog and parasite eggs would have been deposited) became infected.

Thus it seems very probable that the hosts of monogeneans normally become infected at sites quite close to those where deposition of the eggs of the parasites has occurred.

Tactic behaviour

Phototaxis. Monogenean larvae, according to Bychowsky (1957), are generally positively phototactic, at least in the first part of their free existence, and this is supported by the observations of Bovet (1967) on *Diplozoon* and Paling (1969) on *Discocotyle.* Most oncomiracidia have eyes, and except in most polyopisthocotylineans, there are usually two pairs present; the eyes have lenses and are characteristically orientated, an anterior pair being directed postero-laterally, and a posterior pair directed antero-laterally. Most polyopisthocotylineans have a single pair of pigment cups without special lenses. However, the polyopisthocotylineans *Diplozoon* and *Discocotyle,* like the larvae of most of the other monogeneans that have been studied, are positively phototactic, and so the special role of the two pairs of eyes with lenses in the behaviour of monogenean larvae is not obvious. Eyes are completely absent in *Leptocotyle, Acanthocotyle, Dictyocotyle, Gyrodactylus, Sphyranura, Rajonchocotyle,* and *Diclidophora,* but little or nothing is known of the behaviour of their oncomiracidia, and it is perhaps worth adding that these genera represent an extremly wide taxonomic range of monogeneans; it seems that the loss of eyes may be a convergent adaptation to some special form of behaviour. In *Acanthocotyle* and *Sphyranura* cilia as well as eyes are absent, and the larvae passively await the arrival of the specific host (see Kearn, 1967b and Alvey,

1936 respectively); *Gyrodactylus* which is viviparous and is also without cilia, has its own special method of invading new hosts (see later); the other four genera have cilia.

According to Bychowsky (1957), it is only in the first stage of their free-swimming existence that oncomiracidia are positively phototactic; in the later stage they lose this response to light. Bychowsky apparently based his conclusions on a study of mainly dactylogyrids and *Nitzschia sturionis,* but Paling (1969) has found a similar pattern of behaviour in *Discocotyle.* During the first stage the haptors of these oncomiracidia are folded and inactive, and, in this condition, the larva of *Discocotyle* was found to be unable to attach itself even when presented with its normal site of attachment on its specific host fish. In the second phase, according to Bychowsky, the haptor unfolds and the sclerites protrude, and the haptor is then ready to attach the larva to the host. However, the larva of *Entobdella soleae* has been found by Kearn (1967a) to attach itself to the skin of sole first by its anterior glands, and only after this does the haptor unfold.

Presumably a positive phototaxis would nearly always result in the oncomiracidia swimming upwards towards the surface, and while Paling (1969) has claimed that this would bring the larvae of *Discocotyle* to the region where the host trout feed in summer-time, such behaviour in *Diplozoon* would take the larva away from the bottom-feeding host fish bream. Though a positive phototaxis has not been reported for the oncomiracidia of *Entobdella soleae,* these larvae have two pairs of eyes, and if they do indeed continue to swim upwards, they would be taken far above the region where, according to Kearn (1967a), the host fish *Solea solea* spends its time, i.e. either buried in the mud during the day-time, or cruising a few centimetres above it during the night.

The positively phototactic phase lasts only 2-3 min in the majority of dacylogyrids and for up to 10-12 min in *Nitzschia sturionis* (see Bychowsky, 1957), with the succeeding phase lasting much longer: not less than 4-5 h in dactylogyrids, and about 24 h in *Nitzschia.* The relative durations of these phases are however reversed in *Discocotyle,* where Paling (1969) found phase I to last about 1-2½ h, with phase II being much shorter at about ½ h, though with a range of from 8 to 53 min. It is obvious that many more species need to be studied before any worthwhile analysis can be made of the general biological significance of the two phases.

Chemotaxis. In *Entobdella* the oncomiracidium was found by Kearn (1967a) to be attracted to a very highly specific odour in the epidermal mucus of its host *Solea solea,* but when Paling (1969) applied Kearn's experimental techniques in investigating host-finding in *Discocotyle,* he found that there was no positive chemotaxis to the mucus of trout. Neither does there seem to be any chemical factor involved in the response of *Diplozoon* to bream (Bovet, 1967). It appears, however, that some polystomatids may use a chemotaxis in host-finding, e.g. *Protopolystoma xenopi,* as observed by Thurston (1964), invades directly the cloaca of *Xenopus* and then continues up the ureter to the kidney, and the most obvious explanation of this behaviour is that the larva may be exploiting a chemical trail.

Rheotaxis. The larvae of *Polystoma integerrimum* have been seen (Llewellyn, 1957) to make what appeared to be random collisions with frog tadpoles before gliding haphazardly over the skin until by chance they found themselves within a millimetre or so of the edge of the operculum, when their behaviour suddenly changed. The larvae then made for the very edge of the operculum, paused for some seconds, and then swiftly entered the opercular cleft during an interval between the exhalant pulses from the gills. Since the larva chooses a moment of slack water, it is doubtful if this behaviour is a conventional rheotaxis, but it is clearly related to a perception of water currents. There is evidence also of an appreciation of water currents in the invasive behaviour of *Diplozoon.* Here Bovet (1967) discovered that the oncomiracidium swam randomly until it found itself by chance in the cone of water in front of the mouth of the fish. Once in this region its cilia stopped beating if there were currents actually being drawn into the mouth, but the larva started swimming again in slack water. After several such periods when locomotion was inhibited, the oncomiracidium finally entered passively into the mouth of the fish. Thus, as in *Polystoma,* so in *Diplozoon* there is evidence of a perception of water currents.

Behaviour of the larvae after making contact with the host

Having made contact with its host, the oncomiracidium immediately attaches itself, generally (according to Bychowsky, 1957) by using the haptor and its armature, but according to Kearn (1967a) *Entobdella* first attaches itself by secretions from its anterior glands and *then* unfolds its haptor and attaches itself posteriorly.

Immediately after attachment to the host the oncomiracidium of *Entobdella soleae* sheds its ciliated epidermis (Kearn, 1967a), and in *Diplozoon paradoxum* the corresponding process was found by Bovet (1967) to be capable of being triggered-off, under experimental conditions, by strong concentrations of gill mucus or host serum. In *Polystoma integerrimum,* however, the epidermis is almost certainly not shed immediately the larva alights on the body of the tadpole, since the actual entry into the gill chamber (see p. 25) is a swimming movement.

The oncomiracidium does not usually invade its host at the precise site of its future habitat; more often there is a migration from the point of initial contact to the definitive habitat. In *Entobdella soleae* the oncomiracidia establish themselves first on the upper surface of the sole, and only migrate to the under surface at about the time of sexual maturity (Kearn, 1963). Some monogenean gill parasites, after shedding their ciliated epidermis, may reach the gill chamber by crawling or gliding over the surface of the body or the buccal cavity, e.g. *Neodactylogyrus crucifer* on *Rutilus rutilus, Tetraonchus monenteron* on *Esox lucius,* and *Diplectanum aequans* on *Morone labrax* (see Kearn, 1968); others, e.g. *Diplozoon paradoxum* on *Abramis brama* and *Discocotyle sagittata* on *Salmo trutta* (see Bovet, 1967 and Paling, 1969 respectively) appear to retain their cilia until they are carried to the gill chamber by the ventilating currents of their hosts.

Even after reaching the gills, the pattern of invasive behaviour of gill parasites may not have been completed. Paling (1969) has shown that the distribution of *Discocotyle sagittata* on the gills of trout is not that which would be expected from a purely passive invasive process; a post-contact migration almost certainly takes place. While nothing is known of the mechanics of invasion of *Gadus merlangus* by *Diclidophora merlangi,* the marked preference exhibited by the parasite for the anteriormost gill of its host could be the result of a migration performed after initial contact has been made; it is known that some polyopisthocotylineans, which do not move about as adults, have mobile juvenile stages, e.g. the diporpae of *Diplozoon paradoxum.* The migration of *Polystoma integerrimum* from the gills to the bladder of the tadpole is well known, and Combes (1968) has shown that even though the period of development on the gills is not a necessary preliminary to subsequent development in the bladder, nevertheless oncomiracidia placed on metamorphosing tadpoles (i.e. at the time when the juveniles would migrate from the gills to the bladder) still make an excursion to the gills, indicating that the stimulus arises within the parasite, and is not provided by the particular developmental state of the host.

SPECIAL WAYS OF INVASION

The above account refers to the common pattern of invasive behaviour in monogeneans, but there are some exceptions. *Acanthocotyle lobianchi* and *Sphyranura oligorchis,* according to Kearn (1967b) and Alvey (1936) respectively, have non-ciliated oncomiracidia which are presumably non-motile, and the larvae simply wait until a host comes to rest upon them, in the former a species of *Raia* and in the latter, a salamander.

Gyrodactylus is viviparous, and the parasites are almost fully-grown when born and are capable of transferring to other host fishes if these swim sufficiently close to the "parent" host. The parasites themselves seem incapable of swimming, and it is not known through what agency they recognize the proximity of a suitable host fish. It seems unlikely that they make random leaps, since it is known that infection may spread rapidly among hosts confined in aquaria or ponds. From time to time gyrodactylids have been recorded as occurring free on the bottoms of aquaria, and so it is possible that an intermediate stage in the invasive procedure is a resting on the bottom to await the approach of a suitable host. But whatever the means of transfer from host to host, population studies on *Gyrodactylus* have indicated that the majority of the progeny perish (Bychowsky, 1957).

The invasive behaviour of two gastrocotylids *Pricea multae* and *Gotocotyla* sp. involves settling on the gills of non-specific small fishes that are prey for the predatory large specific hosts of the parasites, i.e. an intermediate host is incorporated into the life-cycle (Bychowsky & Nagibina, 1967).

CONCLUSIONS

Perhaps the main conclusion to emerge from the above review is an appreciation of the extreme paucity of our knowledge about the invasive behaviour of monogeneans. Information is available about no more than half-a-dozen species, and it is not known how representative these may be. Any attempt to make generalizations from such meagre information is of course fraught with danger; yet the situation is substantially improved upon that prevailing when a previous attempt was made to theorize (Llewellyn, 1957). The inference drawn at that time from considering the invasive behaviour of

Polystoma integerrimum was that in general gill parasites might invade their hosts by first finding the larger target of the general body surface and then subsequently locating the gills. While recent work has shown that some gill parasites (e.g. *Diplectanum aequans*) in fact do just that, some other gill parasites (e.g. *Discocotyle sagittata* and *Diplozoon paradoxum*) invade the gills by being carried through the mouth with the gill ventilating current.

It seems then that we can expect to find a variety of ways in which monogeneans invade their hosts. Some common features may be that slow-swimming parasite larvae usually make contact with their fast-swimming hosts only when the latter are for some reason stationary near the bottom; that host-location in skin-parasites may be by chemotaxis; that gill parasites attain the gill chamber either via the body and opercular opening or more directly via the mouth; and that since some gastrocotylids have been found to simulate cestodes and (in respect of second intermediate hosts) digeneans in their employment of an intermediate host, some other monogeneans will be found convergently to have incorporated an intermediate host into their life-cycles or to have adopted some other complex pattern of invasive behaviour.

REFERENCES

ALVEY, C. H., 1936. The morphology and development of the monogenetic trematode *Sphyranura oligorchis* (Alvey, 1933) and the description of *Sphyranura polyorchis* n. sp. *Parasitology, 28:* 229-253.

BAINBRIDGE, R., 1960. Speed and stamina in three fish. *J. exp. Biol., 37:* 129-153.

BOVET, J., 1967. Contribution à la morphologie et à la biologie de *Diplozoon paradoxum* v. Nordmann, 1832. *Bull Soc. neuchâtel. Sci. nat., 90:* 63-159.

BYCHOWSKY, B. E., 1957. *Monogenetic trematodes, their classification and phylogeny:* 509 pp. Moscow: Leningrad, Academy of Sciences, U.S.S.R. English translation by W. S. Hargis and P. C. Oustinoff, 1961. Washington: American Institute of Biological Sciences.

BYCHOWSKY, B. E. & NAGIBINA, L. F., 1967. On "intermediate hosts in monogeneans (Monogenoidea)." *Parassitologia, 1:* 117-123 (in Russian).

COMBES, C., 1968. Biologie, écologie et biogéographie de digènes et monogènes d'amphibiens dans l'est des Pyrénées. *Mém. Mus. natn. Hist. nat., Paris (Ser. A. Zool.), 51:* 1-195.

EUZET, L. & RAIBAUT, A., 1960. Le dévelopement postlarvaire de *Squalonchocotyle torpedinis* (Price, 1942) (Monogenea, Hexabothriidae). *Bull. Soc. neuchâtel. Sci. nat., 83:* 101-108.

KEARN, G. C., 1963. The life cycle of the monogenean *Entobdella soleae,* a skin parasite of the common sole. *Parasitology, 53:* 253-263.

KEARN, G. C., 1965. The biology of *Leptocotyle minor,* a skin parasite of the dogfish, *Scyliorhinus canicula. Parasitology, 55:* 473-480.

KEARN, G. C., 1967a. Experiments on host-finding and host-specificity in the monogenean skin parasite *Entobdella soleae. Parasitology, 57:* 585-605.

KEARN, G. C., 1967b. The life cycles and larval development of some acanthocotylids (Monogenea) from Plymouth rays. *Parasitology, 57:* 157-167.

KEARN, G. C., 1968. The development of the adhesive organs of some diplectanid, tetraonchid and dactylogyrid gill parasites (Monogenea). *Parasitology, 58:* 149-163.

KEARN, G. C., 1970. The oncomiracidia of the monocotylid monogeneans *Dictyocotyle coeliaca* and *Calicotyle kroyeri. Parasitology, 61:* 153-160.

LLEWELLYN, J., 1957. Host specificity in monogenetic trematodes. *First Symposium on Host Specificity among parasites of Vertebrates:* 191-212. Neuchâtel.

LLEWELLYN, J., 1962. The life histories and population dynamics of monogenean gill parasites of *Trachurus trachurus. J. mar. biol. Ass. UK., 42:* 587-600.

LLEWELLYN, J. & TULLY, C. M., 1969. A comparison of speciation in diclidophorinean monogenean gill parasites and in their fish hosts. *J. Fish. Res. Bd Can., 26:* 1063-1074.

OWEN, I. L., 1970. The oncomiracidium of the monogenean *Discocotyle sagittata. Parasitology, 61:* 279-292.

PALING, J. E., 1965. The population dynamics of the monogenean gill parasite *Discocotyle sagittata* Leuckart on Windermere trout, *Salmo trutta,* L. *Parasitology, 55:* 667-694.

PALING, J. E., 1969. The manner of infection of trout gills by the monogenean parasite *Discocotyle sagittata. J. Zool., Lond., 159:* 293-309.

PROST, M., 1963. Investigations on the development and pathogenicity of *Dactylogyrus anchoratus* (Duj., 1845) and *D. extensus* Mueller et v. Cleave, 1932 for breeding carps. *Acta parasit. pol., 11:* 17-48.

THURSTON, J. P., 1964. The morphology and life cycle of *Protopolystoma xenopi* Price (Bychowsky) in Uganda. *Parasitology, 54:* 441-450.

WISKIN, M., 1970. The oncomiracidium and post-oncomiracidial development of the hexabothriid monogenean *Rajonchocotyle emarginata. Parasitology, 60:* 457-459.

DISCUSSION

H. D. Chapman

Does the behavioural response you have described for *Diplozoon* imply that some form of mechanical stimulus (rheotaxis?) triggers the cessation of the beat of the cilia and subsequent cessation of movement of the oncomiracidium?

J. Llewellyn

This is the obvious implication, but there is no supporting experimental or even morphological evidence.

R. M. Anderson

You suggested with reference to gill monogeneans that active selection by the larval parasite occurs in relation to a site for attachment. Would you agree that gill water currents which differ markedly in nature, in different species of fish, must also play an important part, if not entirely determining where the parasite attaches to the gills?

J. Llewellyn

As I tried to indicate in my talk, not all gill-dwelling monogeneans enter the gill chamber with the ventilating currents—some invade through the operculum. In those monogeneans which do invade through the mouth, "active selection", if it operates at all, may be imagined as operating either at first contact with the gills, or by migration after contact, or both processes may occur. Dr Paling has shown that the distribution of *Discocotyle sagittata* on trout gills is different from that of glochidia introduced experimentally, and his investigations provide fairly convincing evidence that gill currents are not solely responsible for the distribution of the monogenean. I have made some preliminary observations (*Proc. 1st Int. Congr. Parasitol.,* I: 543-545 (1966)) that *Plectanocotyle gurnardi* may move about on the gills of *Trigla cuculus,* and continuing long-term investigations are so far certainly not disproving this.

   Thus while agreeing that water currents are likely to be found to play an important part in determining the definitive distribution of some monogeneans on the gills of their hosts, there is substantial evidence that other factors may be involved.

K. M. Lyons

With regard to the distribution of monogeneans over the gills of the host fish it would be interesting to know how much the parasites migrate over or even between gills in a post-larval stage. Perhaps the problem is not always as simple as the larvae settling out on regions of the gills most exposed to the ventilating current. I believe that Dr Llewellyn has described what may represent active site selection in that he found adult *Kuhnia scombri* on mackerel tend to prefer the more sheltered regions of the gills and tend to occur towards the proximal ends of the primary lamellae where the two hemibranchs join together.

J. Llewellyn

I would like to thank Dr Lyons for reminding me of the situation in *Kuhnia scombri*; here it seems very likely that the gill ventilating currents of the mackerel may play a large part in determining that it is the first gill which is overwhelmingly the one parasitized in relation to the other three, but correspondingly unlikely that these same currents wash all the larvae into sites near the gill arch sheltered by the inter-branchial septum; it seems exceedingly probable that some migration must take place after initial contact. Such a migration is known to take place in the diporpa larvae of *Diplozoon paradoxum*. Whether or not the migrations may extend over more than one gill remains to be investigated.

Behaviour of larval nematodes

NEIL A. CROLL

Department of Zoology and Applied Entomology
Imperial College of Science and Technology, London

Information on nematode behaviour is largely about aspects of infective larvae readily interpreted in terms of field migrations, host location and host recognition. Taxes have been described from correlations between external stimuli and larval distributions. On behavioural evidence alone, sensitivity has been deducted to: light, heat, chemicals, mechanical stimulation, gravity and electric fields.

Mechanisms controlling larval activity in responding to separate stimulation modalities have been interpreted using controlled levels of stimulation and pharmacological agents. Energy consumption of activity is discussed.

Larvae have been tracked on agar, and the tracks analysed. Tracks of preinfective and infective larvae are compared. Those larval tracks investigated are dominated by reversals and arcs due to persistent asymmetries in successive undulations. These phenomena are discussed as possible orientation mechanisms. Tracks show individual idiosyncrasies, and are influenced by stimulation frequencies. Orientation mechanisms and population behaviour are discussed.

CONTENTS

INTRODUCTION

The study of nematode behaviour is passing from embryogenesis to infancy, but the research effort to date is unbalanced. Of the several hundred reports of locomotory responses, all but a very few have reported directional migrations, which may be interpreted as host location and host recognition mechanisms. Many of the other behavioural performances are poorly known. Investigations have been made on some of the integrated behavioural events in larvae, which require locomotory movements, such as hatching, feeding, moulting, exsheathing, penetrating and moving through tissues. Out of these researches have grown more detailed physiological studies on activity, sensory mechanisms, orientation responses, movement patterns and the energetics of locomotion. The subject has probably gained much because those in it have paid more than lip-service to the benefits of alliance between those studying larval animal and plant parasites and wholly free-living forms (Wallace, 1961a, 1963; Lee, 1965; Klingler, 1965; Van Gundy, 1965a; Croll, 1970a).

Nematodes have a very primitive central nervous system, with bilateral organization and surface receptors of separate modality. A study of the complexity and mechanisms of nematode behaviour, has a wide zoological relevance to students of comparative physiology and ethology.

NEMATODE SENSES

Most orientation behaviour has been described from an analysis of correlations between external stimulation and larval distributions. Inconclusive results may mean only an *apparent* lack of a *directed* response, by larvae of one particular age and physiological history, at the *intensity* levels and *gradient* slopes used. Many orientation responses are completely reversed at different stimulus intensities (Fraenkel & Gunn, 1940) but this has rarely been examined for nematode larvae. Kinetic responses (changes in the rates of activity or the rates of turning) would have remained undetected in most experimentation, and negative results in no way guarantee that a response has not occurred.

Nematode locomotion is influenced by environmental factors. These factors, e.g. temperature and anaerobiosis, may directly affect metabolism or act through integrated receptor-neuromuscular co-ordination. The non-nervous effects typically limit the nervous ones, determing whether activity is possible and its rate following stimulation. On behavioural evidence alone, without any electrophysiological evidence it has been suggested that larval nematodes respond to sensory-mediated stimulation. A few examples of such responses will be given (Croll, 1970a); it will be of interest to compare the senses predicted on behavioural evidence, with the sense organs to be discussed by Dr S. Brenner.

Light

There has often been inadequate separation of heat from light sources, which has made many results somewhat inconclusive (Parker & Haley, 1960; Wallace, 1961a). Nevertheless, in the field, light was found to influence migration of *Haemonchus contortus* (Rees, 1950), and other trichostrongyles (Rogers, 1940). A dermal light sense (Steven, 1963) has been suggested for larval parasites (Croll, 1965, 1966a; Wilson, 1966), which were activated upon illumination following dark adaptation, and showed a photoklinokinesis (Croll, 1965, 1966a) (Fig. 1). Wallace (1961b) was unable to observe a phototaxis using infective *Ditylenchus dipsaci*. Many free-living nematodes are reported as having ocelli, and these function as photoreceptors in *Chromadorina viridis* (Croll, 1970a; Croll, Riding & Smith, unpubl.). It may be noteworthy that in a recent review (Croll, 1970a) 30 investigations were reported, including six positive phototaxes, seven negative phototaxes, six photic indifferences and 11 showing some effect but not taxes. Aside from the possible exception of *Mermis subnigrescens,* no parasitic forms have obvious photoreceptors (Cobb, 1929; Croll, 1966c; Ellenby, 1964).

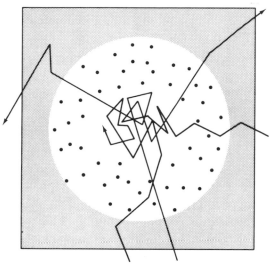

Figure 1. The photoklinokinetic response of infective larval *Trichonema* spp., tracked crossing an illuminated arena. Heavy stippling, 5 c/ft^2, light stippling, 300 c/ft^2; and clear central area, 900 c/ft^2 (Croll, 1965).

Heat

Ancylostoma, Necator and *Strongyloides* spp. infective larvae move up through heat gradients (Khalil, 1922; Fulleborn, 1924, 1932; Lane, 1930, 1933; Reesal, 1951 and others). Thermal convection may have influenced larvae in earlier experiments, but Parker & Haley (1960) reported positive thermotaxes in *Nippostrongylus brasiliensis* infective larvae on agar. The plant parasites *D. dipsaci* and *Pratylenchus penetrans* on agar, moved up gradients of 0.033°C/cm, a comparable thermal gradient to that around germinating seeds (El-Sherif & Mai, 1969). Adult *N. brasiliensis, Camallanus* sp., and *Rhabdias bufonis* (the latter in cold-blooded vertebrates), keep moving up heat gradients until they die (McCue & Thorson, 1964). A striking feature of most nematode thermotaxes is their thermal discrimination, often approaching 0.01°C/cm.

Ronald (1960) with larval *Anisakis decipiens* and Wallace (1961b) with *D. dipsaci* larvae, observed thermal preferenda of 32.5° and 10°C, respectively. The preferendum of *D. dipsaci,* was altered by different pre-storage temperatures (Croll, 1967a) (Fig. 2).

Chemicals

Because of the relative immobility of plants, physical and chemical gradients tend to form in the rhizosphere and this has provided an interesting

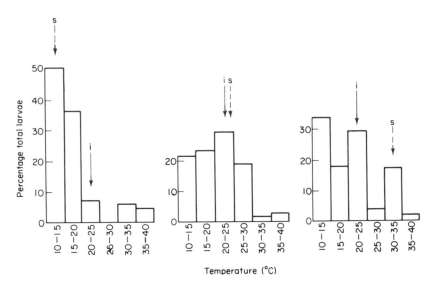

Figure 2. *Ditylenchus dipsaci* infective larvae after 30 days, storage at 10°, 20° and 30° C respectively, and then inoculated into a thermal gradient from 10° to 40° C, and left for 22 h. Acclimatization occurred, larvae showed a preference for the storage temperature ($P < 0.001$) modified from Croll, 1967a). s, Storage temperature; i, inoculation point.

experimental situation regarding nematode behaviour around host roots (Bird, 1959, 1960, 1962; Jones, 1960; Lownsbery & Viglierchio, 1961; Viglierchio, 1961; Blake, 1962; Croll, 1970a). Carbon dioxide has become of central interest; it is attractive to many plant and animal parasitic species (Fig. 3) (Klingler, 1963, 1970; Sasa *et al.*, 1960; Croll & Viglierchio, 1968), and acts as an exsheathment trigger for some trichostrongyles (Rogers, 1966). Carbon dioxide in exhalent air activated and attracted hookworm larvae (Sasa *et al.*, 1960), and the leaf parasites *Aphelenchoides rizemabosi*, and *Aphelenchoides fragariae* penetrated artificial stomata out of which CO_2 was passing (Plate 1) (Klingler, 1970). Chemical stimuli are certainly involved in the control of microfilarial periodicity (Hawking, 1967), sex attraction (Green, 1966), and gustatory stimulation (Roberts & Fairbairn, 1965).

Mechanical

There is ample evidence from direct observations and cine films that the hatching, feeding and penetration of larvae involve tactile sensitivity. An "angle sense" has been reported for some active penetrators (Lane, 1930; Fulleborn, 1932 and others), which orientate at right-angles to a solid surface. The only experimental work is that of Croll & Smith (1970), who stimulated body regions of *Rhabditis* sp. using measured intensities of kinetic energy. The responses varied in different parts of the body and showed central adaptation (Fig. 4). Larvae reported to show rheotaxes may have mechanoreceptors or proprioceptors, but whether these movements are sensorily mediated is unresolved.

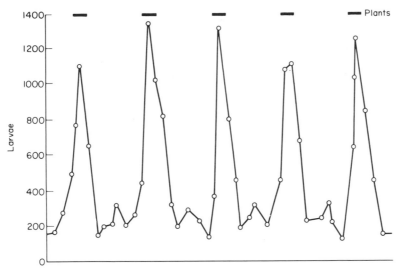

Figure 3. The distribution of *Meloidogyne hapla* infective second stage larvae in sand, five days after being placed under tomato seedlings. The bars indicate the position of the seedlings (redrawn after Viglierchio, 1961).

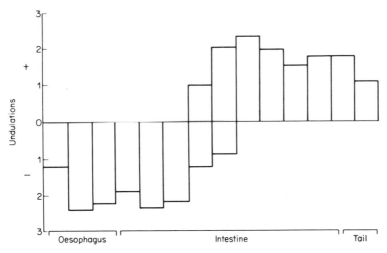

Figure 4. The mean response forwards (+) or backwards (−) when different regions of the body of previously stationary, unadapted adult *Rhabditis* sp. were stimulated with an impact force of 14.5 dynes/cm (Croll & Smith, 1970).

Gravity

Vertical migration of strongyles, trichostrongyles, and hookworms in the field, led to the early assumption that infective larvae respond to gravity (Kates, 1950; Michel, 1969). When spreading randomly from a point source some larvae will go up and Crofton (1954) argued that greater displacement would be accomplished along narrow blades of grass, than over horizontal surfaces. Surface films acting on larvae exert a force 10^4-10^5 times greater than gravity (Crofton, 1954).

These experiments, plus an analysis of "random walk" by Broadbent and Kendall (1953) based on Crofton's data for *Trichostrongylus retortaeformis* infective larvae, put in question the validity of geotaxes of larvae. Crofton's experiments may have been too short as directional preferences become more pronounced with time (Odei, 1969, and below, Fig. 5). Negative geotaxes have been described by many authors including: Africa, (1931); Barraclough & French (1965); Buckley (1940); Cunningham (1956); Odei (1969); and Wallace

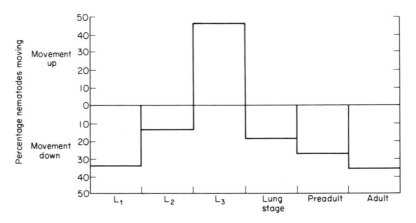

Figure 5. The distribution of different stages of *Nippostrongylus brasiliensis* moving freely in a vertical field. Activity differences are disregarded, but infective larvae may be seen to move against gravity (data from Cunningham, 1956).

(1961b). *Turbatrix aceti* is tail heavy (Peters, 1928), and similar morphological factors, including any drag provided by the sheath, may influence vertical movements. *A. tubaeforme* infective larvae centrifuged at 500 g for 30 min, although less active afterwards, still moved normally on agar (Croll, unpubl.), whereas intense stimulation of other kinds inhibited normal behaviour. The absence of obvious sense organs cannot be serious grounds for scepticism when the shortage of knowledge on nematode sense organs is considered.

Electric fields

The doubt that surrounds the possibility of sensorily-mediated orientation to galvanotaxes is much greater even than that surrounding orientation to gravitational fields. Since it was realized that the potentials developed around plant roots (Etherington & Higginbotham, 1960), are double or treble the 30 mV/mm threshold of some nematode responses, these have been seriously investigated. Bird (1959) argued that larvae were influenced by redox potentials, and moved along a gradient of decreasing potential. All of the dozen genera of plant parasites except second stage *Heterodera schachtii* larvae move to the cathode (Bird, 1959; Caveness & Panzer, 1960; Jones, 1960; Croll 1967b). Larval *T. retortaeformis,* in currents of 10-40 mA moved towards an agar bridge cathode (Gupta, 1962). Microfilariae of *Litomosoides carinii* in standard serum-tyrode (Hawking *et al.,* 1950) and larval *Pelodera strongyloides* moved to the anode, whereas adults did not respond (Whittaker, 1969).

The sensory basis of galvanotaxes if any, is quite obscure, but protozoans, molluscs, goldfish, lampreys and others move directionally in electric potentials (Jahn, 1961; Nikolsky, 1963), and the possibility of nematodes being influenced by electric potentials cannot be overlooked. The immediacy of the response to a change in current direction (Hawking *et al.,* 1950; Jones, 1960), and the size of some larvae responding eliminates the universal explanation that the response is to diffusible chemical products of electrolysis or electro-phoresis.

Other senses and interactions

It has been observed that nematode larvae move up water gradients (Wallace, 1961a), but as they can only move when in a fluid, this seems to be no more than a physical factor. Orientation to magnetic fields has not yet been examined for nematodes, but may exist in many invertebrates (Barnwell & Brown, 1961; Brown, 1969).

The senses listed above have been treated individually whereas complex interactions prevail naturally. Responses to two or more stimulus variables have not yet been seriously studied for nematode larvae.

LOCOMOTORY ACTIVITY

From the data summarized above, it may be deduced that nematodes are sensitive to certain, nervously mediated stimuli. Tactic movements represent only a very limited part of their behavioural repertoire and larval activity is a more basic parameter for measuring nematode behaviour.

Sensory control of larval activity

As a first approximation, it has been suggested that a fundamental contrast exists between the different biological categories of nematode in the bases for their activation (Croll, 1970b). Totally free-living nematodes (*C. viridis, Panagrellus redivivus, Rhabditis* spp., and *T. aceti*), bacteriophagous, pre-infective larvae (*Trichonema* spp. and *A. tubaeforme*) and some plant parasites (*Aphelenchoides* spp., *Aphelenchus avenae* and *D. dipsaci*) all show endo-genously-based spontaneous activity, in the absence of external stimulation. In all these categories there may be an additional response to exogenous stimulation, superimposed on prevailing endogenous activity. This is as expected, for most of these are foraging forms (Croll, 1972a). At the second moult a major change occurs in the pattern of activation of animal parasitic forms. Endogenous activity of infective larvae is minimal and occurs in response to external environmental changes. The activity which results from stimulation is of nearly constant duration and frequency for the infective larvae of any one species and physiological history, being largely independent of the modality of stimulation (Croll, 1970b, Fig. 6). Furthermore the extent of the response is comparable for both short-term mechanical and continuous photic

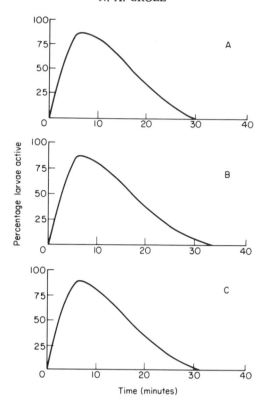

Figure 6. Activity of *Trichonema* spp. infective larvae after a change from ambient conditions. A, Dark adapted 24 h, then illuminated at 400 c/ft²; B, illuminated 24 h, then agitated for 5 min; C, dark adapted and agitated 24 h, then illuminated at 400 c/ft² (modified from Croll, 1970b).

stimulation. It is possible that a change in certain exogenous conditions releases a "packet" of activity, to which the larva is committed. The questions arise: how are the larvae activated?, why do they become inactive following stimulation?, and especially are these phenomena limited by nervous or metabolic factors?

Dark-adapted, infective *A. tubaeforme* larvae were activated upon illumination, but the response became weaker with alternate hourly periods of darkness and light at 20° C. When no longer responding to illumination these larvae were strongly activated by mechanical stimulation (Croll & Al-Hadithi, 1972). In a second series of experiments, larvae which had become inactive following illumination were exposed to an increased light intensity (Figs 7 and 8). Larval activity could be greatly prolonged by a series of five-fold increases in the incident light intensity. Such stimulation resulted only from increases, not decreases of a similar magnitude, and they had to be sudden. Step-wise, gradual increases, even to considerable intensities elicited only low levels of activity. All these changes in light intensity were controlled by neutral density filters, and a constant voltage transformer was included for a constant spectral composition of the light.

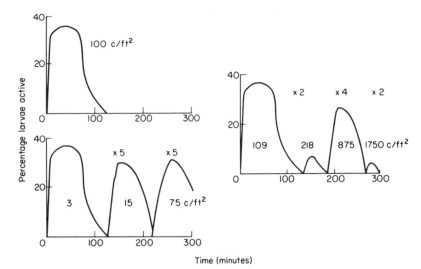

Figure 7. Activity of *Ancylostoma tubaeforme* infective larvae, illuminated following dark adaptation. Initial exposures of 3 c/ft², 100 c/ft² and 109 c/ft² elicited equivalent responses. Following inactivity, further regimes of activity were stimulated by sudden increases of four- and five-fold, independent of the actual intensity (Croll & Al-Hadithi, 1972).

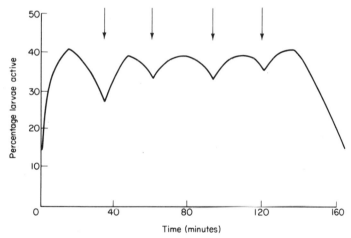

Figure 8. Activity of *Ancylostoma tubaeforme* infective larvae, maintained by successive five-fold increases in light intensity. Following darkness, intensities were 2.8, 14, 350, 1750 c/ft². Arrows show times of increasing light intensity (Croll & Al-Hadithi, 1972).

Pharmacological experiments in larval activity

Nematodes are believed to have cholinergic nerves (Goodwin & Vaughn Williams, 1963; Del Castillo & Morales, 1969), so the effects of pharmacological compounds on activity were tested on suspensions of *A. tubaeforme* infective larvae (Fig. 9). Larvae were activated for varying periods by acetyl choline chloride, d-tubocurarine chloride, 5-hydroxy tryptamine and nicotine. Activity was suppressed by succinyl choline chloride (Castillo & de Beer, 1951) and was unaffected by physostigmine and γ-amino butyric acid. Vigorous

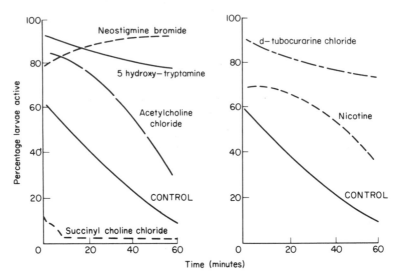

Figure 9. Activity of *Ancylostoma tubaeforme* infective larvae in aqueous suspensions of various pharmacological agents (modified from Croll & Al-Hadithi, 1972).

activity persisted for days in neostigmine bromide (0.1 mg/ml), which inactivates cholinesterase in vertebrates, causing short stimulation, followed by paralysis. By affecting cholinesterase, acetylcholine hydrolysis is suppressed, prolonging its effects; neostigmine also depolarizes muscles directly (Del Castillo & Morales, 1969). *A. tubaeforme* infective larvae remained continually active in neostigmine bromide in the absence of sensory input. Nervous elements were also demonstrated using indoxyl acetate for non-specific esterase, and Gainer's (1969) modification of acetylthiocholine iodide for cholinesterase.

These observations suggest that neither energy shortage, nor negative feedback of metabolic end products cause inactivity of infective larvae. Activity of infective larvae results from sensory input, and inactivity results from sensory adaptation, probably through raising the threshold level of depolarization for synaptic end plate potentials. Sensory adaptation may be peripheral, not central, each modality becoming adapted independently of the others (Croll & Smith, 1970; Croll & Al-Hadithi, 1972).

The interpretation of these phenomena is tempting but is inevitably simplistic; infective stages must conserve energy, while being sensitive to stimuli which may represent a host. Sensitivity must be acute, which would require a rapidly adapting sensory system to each modality. Nevertheless activity, once released, must proceed for a sufficient time to exploit the trigger stimuli. The "packet" of larval activity may be under neuro-endocrine control, while the sense organs rapidly adapt to the change.

Energetics of larval activity

Energetically, infective larvae are closed systems, they do not feed, grow or reproduce, survival and activity being the only energy consuming processes. The levels of unbound neutral lipid (the food reserves) disappear most rapidly

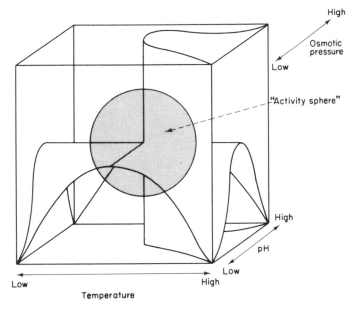

Figure 10. Hypothetical survival curve, described by three physiological parameters known to influence infective larval activity and survival. Each parameter has a low, high and "optimal" value. Outside the cube larvae die, while being quiescent everywhere except the central activity sphere, where activity becomes possible. When quiescent, larvae do not utilize appreciable lipid, in spite of being in "stress" conditions (Croll, 1972b).

under favourable physiological conditions, but in unfavourable conditions activity is lost and *A. tubaeforme* larvae become quiescent (Van Gundy, 1965; Croll, 1972b). Environmental conditions permitting behavioural activity are considerably more restricted than those in which survival is possible (Fig. 10). The QO_2 of active larvae has been found to be only about 5-10% more than that of inactive larvae (Overgaard-Nielsen, 1961), and it has been deduced that activity uses little energy. *A. tubaeforme* in neostigmine bromide consumed lipid at a highly significantly greater rate than those in buffered distilled water (Fig. 11).

The Reynold's number (Taylor, 1951) for infective hookworm larvae is about 10^{-2}.

$$\text{Reynold's number} = \frac{nL^2p}{\mu}$$

(n = frequency of undulations, L = diameter of tail, p = density of medium, μ = viscosity of medium).

This means the inertial stress of movement is less than 1/100th the viscous stress. Motions dominated by viscous forces dissipate energy with the square of the velocity.

$$Pm = A.v^2$$

(Pm = rate of energy dissipation, A = constant dependent on the form of motion, v = the velocity).

The rate of larval movement is therefore the significant factor, and not movement alone as has been reported by many authors. Activity is indeed an

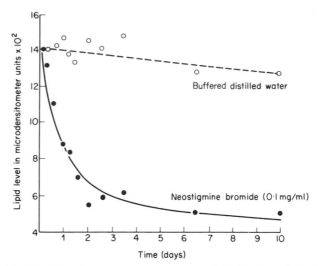

Figure 11. Lipid utilization of *Ancylostoma tubaeforme* infective larvae in buffered distilled water with and without neostigmine bromide (0.1 mg/ml), at 30° C. Neutral unbound lipid in each individual was measured on a scanning microdensitometer at 517 nm after staining in oil red O (Croll, 1972b).

energy utilizing process (Fig. 11), but the conditions in which it is greatest are also those which favour rapid metabolism (Fig. 10).

MOVEMENT PATTERNS

Nematodes inscribe tracks on agar (Sandstedt *et al.,* 1961; Rode & Staar, 1961; Croll, 1971) and in $CaCO_3$ suspensions (Croll, unpubl.) and larvae have been tracked under the influence of various directional and non-directional stimuli (Plates 2 and 3). Measurement of rates of undulation, percentage of larvae active and rates of passage through barriers (Webster, 1964), still neglect information on movement patterns. Complete mathematical analyses are still lacking, but certain features have emerged about the paths followed by larval nematodes.

Moving over flat surfaces larvae do not go in straight lines, nor is the direction of successive undulations random with respect to previous undulations. All the species tracked to date show persistent unilateral biases in their undulations (Figs 12, 14 and 15) (Croll, 1969, 1971). These biases cause a series of arcs to be inscribed in the pattern, each of which may be caused by tens or hundreds of persistent asymmetrical undulations. *Trichonema* spp. infective larvae showed wide variation in tracks even in constant conditions, which somewhat defeated first order interpretations (Croll, 1971).

Forward movements are interrupted by periodic reversing, in which waves pass forward. Reversing is a reflex action occurring spontaneously but its frequency is influenced by external conditions and larval age (Fig. 12). A reversal reflex also accompanies "shock reactions" in many nematode species in response to sudden changes from the ambient (Hawking, 1967; Croll & Smith, 1970; Croll, 1971). *Trichonema* spp. infective larvae produce clusters of successive reversals in bright light (200 c/ft²), the frequency becoming reduced with continuous exposure, suggesting sensory adaptation (Croll, 1971).

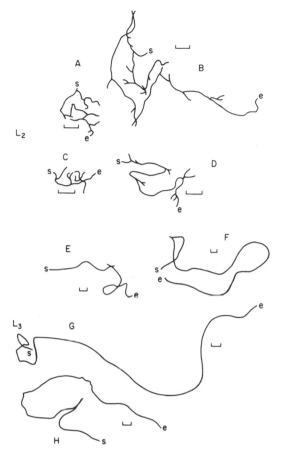

Figure 12. Tracks of *Ancylostoma tubaeforme* larvae on agar, after 15 min at 20° C. A-D, Preinfective L2; E-H, infective L3; s, start; e, end of track. Scale = 1 mm.

The forward direction adopted, immediately following reversals induced by sudden photic stimulation in *Trichonema* spp. infective larvae, was random with respect to the previous direction of movement and compass bearings (Fig. 13) (Croll, 1971). Larvac cxsheathed *in vitro* showed asymmetric patterns and continued to reverse, showing that these phenomena did not result from the presence of a sheath.

Reversal frequency per track length is much greater, and tracks per unit time much shorter in bacteriophagous second stage *A. tubaeforme* larvae, than in infective forms (Fig. 12). These contrasting patterns, separated only by a moult, emphasize that motility without dispersion is necessary in feeding and foraging larvae (Carlson, 1962; Croll, 1969).

Larval dispersal from a point source depends not only on larval activity, but also on the interaction of inherent asymmetries and reversals. The six *selected* tracks of infective *A. tubaeforme* larvae, in gravity and directional light (Fig. 14) may show some of the orientation strategies used. Larvae responding to directional stimulation may reverse to compensate for the disadvantage of inherent asymmetrical bias when it takes them the wrong way (Fig. 14A, also Croll's (1969) interpretation of Green's (1966) data). Similarly it is tempting to

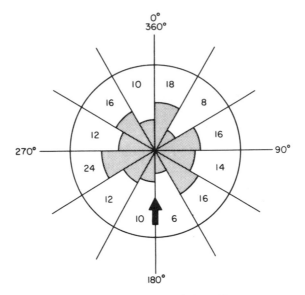

Figure 13. Angles of exit from an illuminated arena followed by 162 *Trichonema* spp. infective larvae. Each larva was strongly illuminated on reaching the centre and reversed before proceeding out of the arena. The arrow indicates the direction of entry of all larvae. There was no preferred quadrat of exit (χ^2 $P = 0.975$) (Croll, 1971).

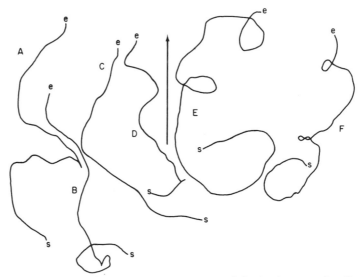

Figure 14. Selected tracks of *Ancylostoma tubaeforme* infective larvae under directional stimulation (acting down the page). A-C, Negative geotaxis; D-F, positive phototaxis; s, start; e, end of track.

suggest that *A. tubaeforme* in light, tighten or close these spirals leading away from the source of stimulation (Fig. 14E, F). Larvae may show confused initial movements, followed by a more persistent directional pattern, towards the stimulus (Fig. 14B); or a steadily increasing accuracy of orientation (Fig. 14C). An initial period of "direction finding" is reported by Odei (1969) with infective *Trichonema* spp. larvae responding to gravity. This may lead to a misinterpretation of results if larvae are left for insufficient periods (Crofton, 1954).

Chandler (1932) observed *N. brasiliensis* larvae remaining in their sheaths, alternately extending and withdrawing. The partial exsheathment of *A. tubae-forme* larvae in drying habitats (Plate 4) "anchors" many larvae to the position of exsheathment. This is an important factor in movement patterns and larval distribution, resulting in a bristling larval population on the drying perimeter of a faecal mass.

Another complication, requiring incorporation into these complex systems, is the tendency for each individual to inscribe its own idiosyncratic path in repeated runs (Fig. 15, Croll, 1971).

The path of an individual *Trichonema* spp. infective larva, tracked at different rates, became more complex until it approached the condition in continuous light at about 50-75 ms intervals (Fig. 16, Croll, 1971). This may suggest a flicker frequency of about 50-75 ms. *P. redivivus* and *C. viridis* orientated to pulsating stimuli of 100 ms and less (Croll, 1967b).

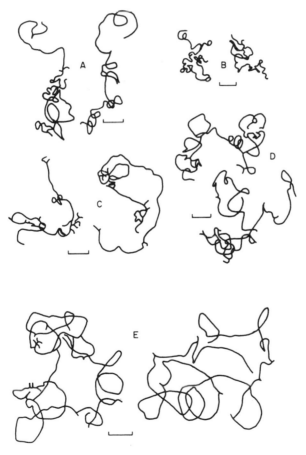

Figure 15. Movement patterns of five pairs of *Trichonema* spp. infective larvae, each tracked twice, to show the idiosyncratic patterns which sometimes occur. Each pair has been drawn for easy comparison, but the compass directions actually differed. A, D, 6.5 c/ft² ; B, C, 200 c/ft² ; E, darkness. Scale = 1mm.

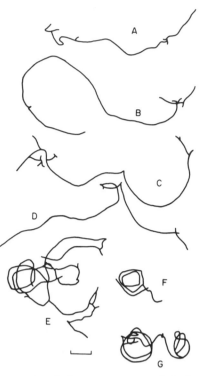

Figure 16. Movement patterns of a single *Trichonema* sp. infective larva in light of constant flash duration (10 μs), but variable flash intervals, 30 min at 20° C. An increasing complexity of the track occurs with decreasing flash intervals, approaching the condition in continuous light (G). Flash intervals: A, 600 ms; B, 300 ms; C, 200 ms; D, 120 ms; E, 100 ms; F, 75 ms (Croll, 1971, except for C, which is corrected from the earlier paper).

Movement patterns provide a useful level, so far hardly examined, for understanding nematode behaviour. Already it is thought that asymmetries and reversals dominate larval locomotion. Idiosyncratic phenomena, and patterns under pulsating stimulation suggest techniques for geneticists and sensory physiologists.

MECHANISM OF ORIENTATION

Nematodes move using longitudinal dorso-ventral muscular hemicylinders, and larvae swim "on their sides" if on two-dimensional surfaces. Paired lateral sense organs are functionally dorso-ventral in movement (Fig. 17). This apparent anomaly has been overlooked by the assumption that orientation is klinotactic (Jones, 1960; Blake, 1962; Klingler, 1963), in which by definition, the environment is sampled alternately and successive stimuli compared.

Another dispute particularly well entrenched in plant nematology is the "apparent" taxis which actually results from a kinetic change in behaviour. Thus it has been argued that larvae moving at random, may become inactivated by chemicals around the root surface. To assume a taxis has occurred, on the final distribution of larvae alone, is to oversimplify the situation. These questions are unresolved, but there seems sufficient data to assume that taxes

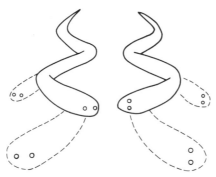

Figure 17. Diagram to demonstrate the problem of bilaterial symmetry in nematode sensory systems. The position of "lateral" sense organs such as amphids and ocelli in the morbid anatomy (left) and the actual position (right) when a nematode moves over a flat surface on its "side".

as well as klinokineses (defined by Fraenkel & Gunn, 1940; Ewer & Bursell, 1950), occur in larval nematodes.

ASSOCIATIONS WITH NON-HOSTS

Dictyocaulus viviparus larvae are translocated on to pasture by the coprophilic phycomycete, *Pilobolus kleinii* (Robinson *et al.,* 1962) (Plate 5). The larvae are activated by light, and may be projected up to 10 f. *Psychoda* carries *Oesophagostomum* and *Ostertagia* larvae but behavioural interactions await investigation (Jacobs, D. E., pers. comm.; Tod *et al.,* 1971).

POPULATION BEHAVIOUR

Some nematode larvae form motile swarms, and immotile aggregations. In such population behaviour, the relative contributions of larvae responding to each other and of physical forces such as water films, are not clear. Nevertheless using microcinematography the dominance of surface forces in synchronous swarming of a *Panagrellus redivivus* population was demonstrated (Croll, 1970c).

SUMMARY

The behaviour of nematode larvae includes hatching, feeding, migration and penetration of tissues, but only the migration of infective larvae has had serious attention to date. Behavioural differences between preinfective and infective larvae can be as characteristic as morphological differences. While much scope exists for more critical work on orientation responses, analyses are now proceeding towards an understanding of the bases of activity and movement patterns. The relationship between activity, quiescence and energy utilization has been discussed. Evidence has been reported for cholinergic nervous transmission, sensory adaptation and thermal acclimatization in larval nematodes. Movement of some larvae is dominated by asymmetrical biases in undulations and reversals and it has been suggested that these may interact to give directional movements.

ACKNOWLEDGEMENTS

I acknowledge the enormous help of Mr J. M. Smith in my work on nematode behaviour. The generous support of the late Prof. B. G. Peters and Prof. T. R. E. Southwood, and of Dr I. L. Riding and the financial opportunities provided by the Royal Society and Medical Research Council of Great Britain are very much appreciated. For permission to reproduce figures and plates I am grateful to: Academic Press (Fig. 6); Allen Press (Fig. 4); Brill (Fig. 2 and Plate 1); and Cambridge University Press (Figs 1, 7, 8, 10, 11, 13 and 16).

REFERENCES

AFRICA, C. M., 1931. Studies on the activity of the infective larvae of the rat strongylid, *Nippostrongylus muris. J. Parasit., 17:* 196-206.

BARNWELL, F. H. & BROWN, F. A., Jr., 1961. Magnetic and photic responses in snails. *Experimentia, 17:* 513-515.

BARRACLOUGH, R. M. & FRENCH, N., 1965. Observations on the orientation of *Aphelenchoides ritzemabosi* (Schwartz). *Nematologica, 11:* 199-206.

BIRD, A. F., 1959. The attractiveness of roots to the plant parasitic nematodes *Meloidogyne javanica* and *M. hapla. Nematologica, 4:* 322-335.

BIRD, A. F., 1960. Additional notes on the attractiveness of roots to plant parasitic nematodes. *Nematologica, 5:* 217.

BIRD, A. F., 1962. Orientation of the larvae of *Meloidogyne javanica* relative to roots. *Nematologica, 8:* 275-287.

BLAKE, C. D., 1962. Some observations on the orientation of *Ditylenchus dipsaci* and invasion of oat seedlings. *Nematologica, 8:* 177-192.

BROADBENT, S. R. & KENDALL, D. G., 1953. The random walk of *Trichostrongylus retortaeformis. Biometrics, 9:* 460-466.

BROWN, F. A., Jr., 1965. How animals respond to magnetism. *Discovery,* November.

BUCKLEY, J. J. C., 1940. Observations on the vertical migration of infective larvae of certain bursate nematodes. *J. Helminth., 18:* 173-182.

CARLSON, F. S., 1962. A theory of survival value of motility. In D. W. Bishop (Ed.), *Spermatozoan motility,* A.A.A.S. Symposium, *72:* 137-146. Washington D.C.

CASTILLO, J. C. & DE BEER, E. J., 1951. The neuromuscular blocking action of succinylcholine (diacetylcholine). *J. Pharmacol. exp. Ther., 99:* 458-464.

CAVENESS, F. B. & PANZER, J., 1960. Nemic galvanotaxes. *Proc. helminth. Soc. Wash., 27:* 73-74.

CHANDLER, A. C., 1932. Experiments on resistance of rats to super-infection with the nematode, *Nippostrongylus muris. Am. J. Hyg., 16:* 750-782.

COBB, N. A., 1929. The chromatropism of *Mermis subnigrescens,* a nemic parasite of grasshoppers. *J. Wash. Acad. Sci., 19:* 159-166.

CROFTON, H. D., 1954. The vertical migration of infective larvae of strongyloid nematodes. *J. Helminth., 28:* 35-52.

CROLL, N. A., 1965. The klinokinetic behaviour of *Trichonema* larvae in light. *Parasitology, 55:* 579-582.

CROLL, N. A., 1966a. Activity and the orthokinetic response of larval *Trichonema* to light. *Parasitology, 56:* 307-312.

CROLL, N. A., 1966b. The phototactic response and spectral sensitivity of *Chromadorina viridis* (Nematoda, Chromadorida) with a note on the nature of the paired pigment spots. *Nematologica, 12:* 610-614.

CROLL, N. A., 1966c. A contribution to the light sensitivity of the *Mermis subnigrescens* chromatrope. *J. Helminth., 40:* 33-38.

CROLL, N. A., 1967a. Acclimatization in the eccritic thermal response of *Ditylenchus dipsaci. Nematologica, 13:* 385-389.

CROLL, N. A., 1967b. The mechanism of orientation in nematodes. *Nematologica, 13:* 17-22.

CROLL, N. A., 1969. Asymmetry in nematode movement patterns and its possible significance in orientation. *Nematologica, 15:* 389-394.

CROLL, N. A., 1970a. *The behaviour of nematodes, their activity, senses and responses.* London: Edward Arnold.

CROLL, N. A., 1970b. Sensory basis of activation in nematodes. *Expl Parasit., 27:* 350-356.

CROLL, N. A., 1970c. An analysis of swarming in *Panagrellus redivivus. Nematologica, 16:* 382-386.

CROLL, N. A., 1971. Movement patterns and photosensitivity of *Trichonema* spp. infective larvae in non-directional light. *Parasitology, 62:* 467-478.

CROLL, N. A., 1972a. Feeding and lipid synthesis of pre-infective *Ancylostoma tubaeforme* larvae. *Parasitology, 64:* 355-368.

CROLL, N. A., 1972b. Lipid utilization and activity of infective *Ancylostoma tubaeforme* infective larvae. *Parasitology, 64:* 369-376.

CROLL, N. A. & AL-HADITHI, I. A. W., 1972. Sensory basis of activity in infective *Ancylostoma tubaeforme* larvae. *Parasitology, 64:* 279-291.

CROLL, N. A. & SMITH, J. M., 1970. The sensitivity and responses of *Rhabditis* sp. to peripheral mechanical stimulation. *Proc. helminth. Soc. Wash., 37:* 1-5.

CROLL, N. A. & VIGLIERCHIO, D. R., 1969. Reversible inhibition of chemosensitivity in a phytoparasitic nematode. *J. Parasit., 55:* 895-896.

CUNNINGHAM, F. C., 1965. A comparative study of tropisms of *Nippostrongylus muris* (Nematoda : Trichostrongylidae). *Biol. Stud. Cath. Univ. Amer., 36:* 1-24.

DEL CASTILLO, J. & MORALES T., 1969. Electrophysiological experiments in *Ascaris lumbricoides*. In G. A. Kerkut (Ed.), *Experiments in Physiology and Biochemistry, 2:* 209-273. London: Academic Press.

EL-SHERIF, M. & MAI, W. F., 1969. Thermotactic response of some plant parasitic nematodes. *J. Nematology, 1:* 43-48.

ELLENBY, C., 1964. Haemoglobin in the chromotrope of an insect parasitic nematode. *Nature, Lond., 202:* 615-616.

ETHERTON, B. & HIGGINBOTHAM, N., 1960. Transmembrane potential measurements of cells of higher plants as related to salt uptake. *Science, N.Y., 131:* 409-410.

EWER, D. W. & BURSELL, E., 1950. A note on the classification of elementary behaviour patterns. *Behaviour, 3:* 40-47.

FRAENKEL, G. S. & GUNN, D. L., 1940. *The orientation of animals.* New York: Oxford University Press.

FULLEBORN, F., 1924. Ueber taxis (tropismus) bei Strongyloides und ankylostomenlarven. *Arch. Schiffs U. Tropenhya., 28:* 144-165

FULLEBORN, F., 1932. Ueber die Taxen und des sonstige Verhalten der infektionsfähigen Larven von *Strongyloides* und *Ancylostoma*. *Z. Bakt. Abt. 1, Originale, 126:* 161-180.

GAINER, H., 1969. Multiple innervation of fish skeletal muscle. In G. A. Kerkut (Ed.), *Experiments in physiology and biochemistry, 2:* 191-208. London: Academic Press.

GOODWIN, L. C. & VAUGHN WILLIAMS, E. M., 1963. Inhibition and neuromuscular paralysis in *Ascaris lumbricoides*. *J. Physiol., Lond., 168:* 857-867.

GREEN, C. D., 1966. Orientation of males *Heterodera rostochiensis* Woll. and *H. schachtii* Schm to their females. *Ann. appl. Biol., 58:* 327-339.

GUPTA, S. P., 1962. Galvanotactic reaction of infective larvae of *Trichostrongylus retortaeformis*. *Exptl. Parasit., 12:* 118-119.

HALEY, A. J., 1962. Biology of the rat nematode, *Nippostrongylus brasiliensis* (Travassos, 1914). II. Preparasitic stages and development in the laboratory rat. *J. Parasit., 48:* 13-23.

HAWKING, F., 1967. The 24-hour periodicity of microfilariae. Biological mechanisms responsible for its production and control. *Proc. Roy. Soc. (B), 169:* 59-76.

HAWKING, F., SEWELL, P. & THURSTON, J. P., 1950. The mode of action of hetrazan on filarial worms. *Br. J. Pharmac. Chemother., 5:* 217-238.

JAHN, T. L., 1961. The mechanism of ciliary movement. 1. Ciliary reversal and activation by electric currents. Ludloff phenomenon in terms of core and volume conductors. *J. Protozoology., 8:* 369-380.

JONES, F. G. W., 1960. Some observations and reflections on host finding by plant nematodes. *Meded. Landb. Hoogesch. Gent., 25:* 1009-1024.

KATES, K. C., 1950. Survival on pasture of free-living stages of some common gastrointestinal nematodes of sheep. *Proc. helminth. Soc. Wash., 17:* 39-58.

KHALIL, M., 1922. Thermotropism in ancylostome larvae. *Proc. R. Soc. Med., 15:* 6-8.

KLINGLER, J., 1963. Die Orientierung von *Ditylenchus dipsaci* in gemessenen Künstlichen und biologischen CO_2 Gradienten. *Nematologica, 9:* 185-199.

KLINGLER, J., 1965. On the orientation of plant nematodes and of some other soil animals. *Nematologica, 11:* 4-18.

KLINGLER, J., 1970. The reaction of *Aphelenchoides fragariae* to slit-like micro-openings and to stomatal diffusion gases. *Nematologica, 16:* 417-422.

LANE, C., 1930. Behaviour of infective hookworm larvae. *Ann. trop. Med. Parasit., 24:* 411-421.

LANE, C., 1933. The taxes of infective hookworm larvae. *Ann. trop. Med. Parasit., 27:* 237-250.

LEE, D. L., 1965. *The physiology of nematodes.* London: Oliver & Boyd.

LOWNSBERY, B. F. & VIGLIERCHIO, D. R., 1960. Mechanism of accumulation of *Meloidogyne incognita acrita* around tomato seedlings. *Phytopathology, 50:* 178-179.

McCUE, J. F. & THORSON, R. E., 1964. Behaviour of parasitic stages of helminths in a thermal gradient. *J. Parasit., 50:* 67-71.

MICHEL, J. F., 1969. The epidemiology and control of some nematode infections of grazing animals. In B. Dawes (Ed.), *Advances in parasitology, 7:* 211-282. London : Academic Press.

NIKOLSKY, G. N., 1963. *The ecology of fishes.* London: Academic Press.

OVERGAARD NEILSEN, C., 1961. Respiratory metabolism of some populations of Enchytraeid worms and free living nematodes. *Oikos, 12:* 17-35.

PARKER, J. C. & HALEY, A. J., 1960. Phototactic and thermotactic responses of the filariform larvae of the rat nematode, *Nippostrongylus muris. Expl Parasit., 9:* 92-97.

PETERS, B. G., 1928. On the bionomics of the vinegar eelworm. *J. Helminth., 6:* 1-38.

REES, G., 1950. Observations on the vertical migrations of the third-stage larvae of *Haemonchus contortus* (Rud) on experimental plots of *Lolium perenne* in relation to meterological and micrometerological factors. *Parasitology, 40:* 127-143.

REESAL, M. R., 1951. Observations on the biology of the infective larvae of *Strongyloides agouti. Can. J. Zool., 29:* 109-115.

ROBERTS, L. S. & FAIRBAIRN, D., 1965. Metabolic studies on adult *Nippostrongylus brasiliensis* (Nematoda : Trichostrongyloidea). *J. Parasit., 51:* 129-138.

ROBINSON, J., POYNTER, D. & TERRY, R. J., 1962. The role of the fungus *Pilobolus,* in the spread of the infective larvae of *Dictyocaulus viviparus. Parasitology, 52:* 17-18.

RODE, H. & STAAR, G., 1961. Die photographische darstellung der kriechspuren (Ichnogramme) von nematoden und ihre loedeutung. *Nematologica, 6:* 266-271.

ROGERS, W. P., 1940. The effect of environmental conditions on the accessibility of third stage trichostrongyle larvae to grazing animals. *Parasitology, 32:* 208-226.

ROGERS, W. P., 1966. The reversible inhibition of exsheathment in some parasitic nematodes. *Comp. Biochem. Physiol., 17:* 1103-1110.

RONALD, K., 1960. The effects of physical stimuli on the larvae of *Terranova decipiens.* 1. Temperature. *Can. J. Zool., 38:* 623-642.

SANDSTEDT, R., SULLIVAN, T. & SCHUSTER, M. L., 1961. Nematode tracks in the study of movement of *Meloidogyne incognita incognita Nematologica, 6:* 261-265.

SASA, M., SHIRASAKA, R., TANAKA, H., MIURA, A., YAMAMOTO. H. & KATAHIRA, K., 1960. Observation on the behaviour of infective larvae of hookworms and related nematode parasites, with notes on the effect or carbon dioxide in the breath as the stimulant. *Japan. J. Exp. Med., 30:* 433-447.

STEVEN, D. M., 1963. The dermal light sense. *Biol. Rev., 38:* 204-240.

TAYLOR, SIR GEOFFREY, 1951. Analysis of the swimming of microscopic organisms. *Proc. R. Soc. (A), 209:* 447-461.

TOD, M. E., JACOBS, D. E. & DUNN, A. M., 1971. Mechanisms for the dispersion of parasitic nematode larvae. 1. Psychodid flies as transport hosts. *J. Helminth., 45:* 133-137.

VAN GUNDY, S. D., 1965a. Nematode behaviour. *Nematologica, 11:* 19-32.

VAN GUNDY, S. D., 1965b. Factors in survival of nematodes. *Ann. Rev. Phytopath., 3:* 43-68.

VIGLIERCHIO, D. R., 1961. Attraction of parasitic nematodes by plant root emanations. *Phytopathology, 51:* 136-142.

WALLACE, H. R., 1960. Observations on the behaviour of *Aphelenchoides ritzemabosi* in chrysanthemum leaves. *Nematologica, 5:* 315-322.

WALLACE, H. R., 1961a. The bionomics of the free-living stages of zoo-parasitic and phyto-parasitic nematodes—a critical survey. *Helminth. Abs., 30:* 1-22.

WALLACE, H. R., 1961b. The orientation of *Ditylenchus dipsaci* to physical stimuli. *Nematologica, 6:* 222-236.

WALLACE, H. R., 1963. *The biology of plant parasitic nematodes.* London: Edward Arnold.

WALLACE, H. R., 1966. Factors influencing the infectivity of plant parasitic nematodes. *Proc. Roy. Soc. (B), 164:* 592-614.

WALLACE, H. R., 1968. The dynamics of nematode movement. *Ann. rev. Phytopathol., 6:* 91-114.

WEBSTER, J. M., 1964. The effect of storage conditions on the infectivity of narcissus stem eelworms. *Nature, Lond., 202:* 571-575.

WHITTAKER, F. W., 1969. Galvanotaxis of *Pelodera strongyloides* (Nematoda : Rhabditidae). *Proc. helminth. Soc. Wash., 36:* 40-42.

WILSON, P. A. G., 1966. Light sense in nematodes. *Science, N.Y., 151:* 337-338.

DISCUSSION

D. L. Lee

I was interested in your work with neostigmine, nicotine etc. and your comments that they may affect neuro-muscular transmission or the contraction of muscles in nematodes. A lot will depend upon the permeability of the cuticle as these third stage larvae do not feed. These compounds could, in fact, affect the sense organs of the nematode rather than having a direct effect upon the muscles or the nerve-muscle junction.

N. A. Croll

It is a long way from isolated preparations of rat muscle to whole nematodes, and the interpretation must be somewhat tentative. Histamine, adrenalin, physostigmine and γ-amino butyric acid all had no observable effect, and nothing was concluded from these or other negative results. When an effect was observed it was concluded that this was due to the chemical agent. The effect could be peripheral at sense organ level, rather than in nerve-nerve or nerve-muscle junctions. I tend to think of a more central action because if the drugs acted peripherally, some form of central nervous fatigue or adaptation would probably modulate the peripheral stimulation after a short period (see Fig. 7). However, some of the effects, especially those of neostigmine bromide and succinyl chloride last for many hours or days.

P. S. Meadows

I was interested in your three-dimensional graph relating activity to environmental conditions. The diagram showed a sphere of maximum activity at the optimum environment. Many free living invertebrates show another smaller sphere within this where animals are inactive because this is their preferred environment. The inactivity of woodlice in high humidities is an example. Do nematodes show this behaviour?

N. A. Croll

You will appreciate that such a graph is very generalized, the sphere may be an ellipse, a rectangle or more likely a "blob". The essential feature of it is to stress the relationship between activity and quiescence, with respect to the environment. As far as my real data go, a plateau does exist for temperature and osmotic pressure in the optimal area, there is no indication however of a decrease in activity in the presence of sensory input. Comparison with woodlice is perhaps a little unfortunate, for infective hookworm larvae probably capitalize on favourable conditions to migrate or penetrate having been under rather different selective pressures than have woodlice.

W. M. Bethel

In your investigations on the effects of external stimuli, particularly with respect to activity, have you ever been able to look at the effects of changes in barometric pressure?

N. A. Croll

I have not related activity to barometric conditions. My colleague Dr Kamala de Soyza however, studied the activity of *Ascaris lumbricoides* larvae and *Heterodera rostochiensis* larvae under greatly increased and reduced pressures. Her unpublished conclusions were that very great differences in pressure, certainly far greater than normal atmospheric fluctuations, were needed before any effect was observed.

R. M. Anderson

You mentioned a start and an end point in the responses of larval nematodes to external stimuli. Have you compared these plots of the nematode tracks, in particular the end point, with simulated plots using theoretical "random walk" models?

N. A. Croll

No, although some analyses are under way. I believe that larvae on agar, move from a point source in random compass directions, in the absence of directional stimuli. The distance that larvae travel, their reversal rates per unit length of track, the asymmetrical arcs they inscribe, and the period of time they are active, are influenced by age and environmental conditions. Because the form of the track is so dependent on conditions I would speculate that a little more than the initial direction of embarkation would be simulated by "random walk" models.

EXPLANATION OF PLATES

PLATE 1

The leaf parasitic nematode *Aphelenchoides fragariae,* moving into artificial stomata, 10-15 μm across, out of which CO_2 was slowly moving at about 2 ml/h (courtesy of J. Klingler).

PLATE 2

Tracks left by *Trichonema* sp. larva moving through a $CaCO_3$ suspension, then dried quickly at 80°C.

PLATE 3

Track of *Ancylostoma tubaeforme* larva moving over 1% ion agar moving in darkness, showing an arc resulting from asymmetrical undulations, and a reversal period (courtesy of J. M. Smith).

PLATE 4

Successive postures of partially exsheathed *Ancylostoma tubaeforme* larva nictating upwards from its sheath, without emerging (courtesy of J. M. Smith).

PLATE 5

Dictyocaulus viviparus infective larvae on the subsporangial swelling of the fungus *Pilobolus kleinii* (black sporangium on top, about 1 mm across) (courtesy of J. Robinson).

Plate 1

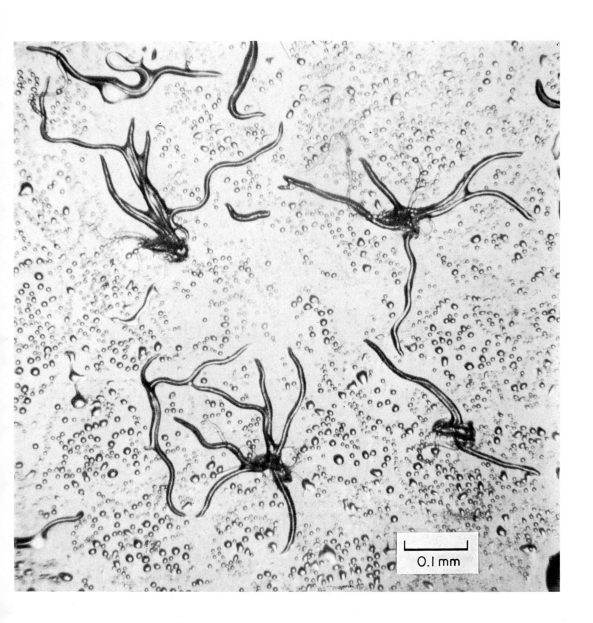

(*Facing p.* 52)

Plate 2

1.0 mm

N. A. CROLL

Plate 3

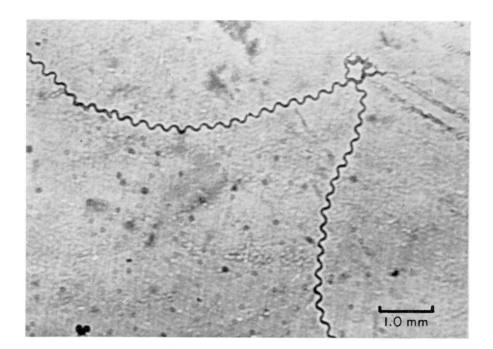

N. A. CROLL

Plate 4

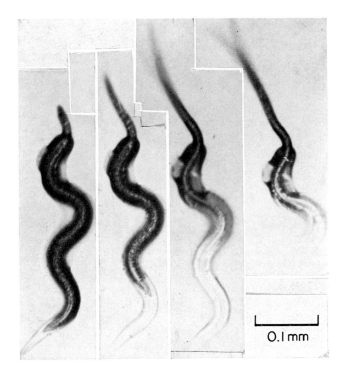

N. A. CROLL

Plate 5

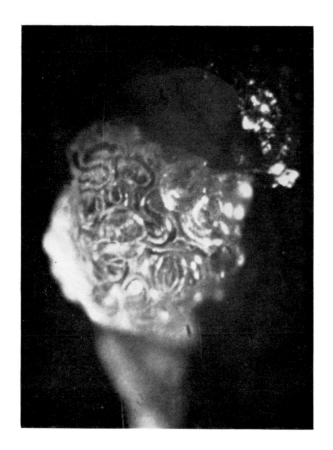

N. A. CROLL

Circadian and seasonal rhythms in blood parasites

M. J. WORMS

National Institute for Medical Research, London

Blood parasites though largely unrelated taxonomically have in common that they all require to be removed from the host's bloodstream for their transmission. Rhythmic variation both on a circadian and a seasonal basis has been found to occur either in the number of parasites present in the blood or in the parasites' reproductive state. These rhythms are particularly well developed among the microfilariae, *Plasmodium* spp. and trypanosomes and may occur in all blood parasites. The nature and timing of these cycles is a feature characteristic of each species or strain of parasite. The means of production and the control of rhythmic cycles also appears to vary between parasites. The significance of the development of rhythmic behaviour by various groups of blood parasites is discussed and the thesis advanced that the cycles may have been brought about by selection pressure from the vector and that survival of the parasite depends upon the maintenance of these cycles.

CONTENTS

INTRODUCTION

Those organisms which at some stage of their life cycle inhabit the blood of vertebrates include the *Haemosporina*, the trypanosomes and the microfilariae. Biologically however, these parasites have a common requirement in that for their transmission and for the continuation of their development they must be taken up from the peripheral blood of the vertebrate host by a blood feeding vector. The number of parasites present in the peripheral blood of the host varies throughout the course of an infection; some variations occur rhythmically and it is these and their relationship to transmission which form the subject of this paper. Rhythms of two lengths have been recognized so far, those which have an approximately 24-h basis (circadian) and those of a much longer period which have been broadly correlated with a yearly cycle termed "seasonal" periodicities.

CIRCADIAN OR 24-H RHYTHMS

Periodicity of microfilariae

The best known and most studied example of a parasite rhythm with a 24-h cycle is the periodicity of microfilariae, the blood-dwelling larvae of some species of filarial nematodes. The phenomenon was first described by Manson in 1879 (see Alcock & Manson Bahr, 1927), who found that the microfilariae of *Wuchereria bancrofti* were absent from the blood of man during the day but appeared there each evening, increasing in number to a peak at about midnight and thereafter decreasing in number until once again they were absent during the day. Periodicity is now known to be characteristic of a number of filarial species parasitizing a wide range of vertebrate hosts including mammals, birds and amphibians (Hawking, 1962; Marinkelle, 1970).

The periodic cycle of many of these species differs from that of *W. bancrofti* both in phase, i.e. whether the maximum number of parasites is present at night (nocturnal) or during the day (diurnal) and in completeness, i.e. the difference between the numbers of parasites present at the peak and the low point of the cycle. In nocturnally periodic *W. bancrofti* for example, this difference is almost 100%, whereas in *Dirofilaria immitis* in the dog, although a definite peak occurs during the evening, approximately 30% of this number is present throughout the whole cycle; this state is defined as sub-periodic (WHO, 1967).

The differences in the nature of the periodic cycle may occur not only between species of filariae but between "varieties" within a single species. Thus *Loa loa* exists in a diurnally periodic form in man, and in a nocturnally periodic form in monkeys (Duke & Wijers, 1958). *W. bancrofti* and *Brugia malayi* each occur in two forms within a single species of host: throughout the major part of its geographic range, *W. bancrofti* is nocturnally periodic but in some areas, particularly in the South Pacific Region, a diurnally subperiodic form is found; similarly *B. malayi* is nocturnally periodic in man for much of its range, but in Malaysia a nocturnally subperiodic form exists, which parasitizes both man and a variety of wild mammalian hosts (Edeson & Wilson, 1964). The complex periodicities reported for *Dipetalonema reconditum* in dogs in different geographic areas by Newton & Wright (1956), Gubler (1966) and Pennington & Phelps (1969) suggest that different varieties of this species may exist also.

The nature of the periodic cycle of some species may be influenced by the species of host in which the parasite occurs, but this influence appears to be confined to those species in which a complete periodicity is not established. Duke & Wijers (1958) showed that the diurnally periodic strain of *L. loa* from man could be transmitted to monkeys in which it retained its diurnal periodicity. Similarly the nocturnally periodic strain of *B. malayi* retains its nocturnal periodicity in monkeys and cats but the nocturnally subperiodic form exhibits a nocturnal subperiodicity in cats but becomes nocturnally periodic when transferred to monkeys (Laing, 1961). These observations may indicate that complete periodicity represents a more highly specialized, stable and hence less plastic state than that of subperiodicity, a concept supported by the narrow host range of *W. bancrofti*, *L. loa*, and the nocturnally periodic form of *B. malayi* which contrasts with the wide range of natural hosts of subperiodic *B. malayi.*

Despite these variations however all filarial periodicities so far studied have had a cycle length of approximately 24 h and, once periodicity is established in a given species of host, it appears to be stable. The periodicity of *Dipetalonema setariosum,* a parasite of mongooses, is shown in Figs 1 and 2. It may be seen that the cycle is of similar nature both in different individuals and during successive parasitaemias in a single individual spanning a period of three years. A similar stability of the cycle of *W. bancrofti* over a period of eight years was reported by Low *et al.* (1933).

These cycles are stable under "natural" conditions in a given locality but, as was first shown by Mackenzie (1882) for *W. bancrofti,* the timing of the cycle depends not upon the presence or absence of light, but upon the activity of the host in relation to the day-night cycle. By causing the host to sleep by day and be active at night for a period of several days the cycle can be reversed. This has been demonstrated many times both for *W. bancrofti* and for other filariae in mammals (Hawking *et al.,* 1963) and birds (Boughton *et al.,* 1938). This reversal depends upon the complete alteration of the activity cycle of the host and not upon any particular component such as feeding, exercise or brief periods of sleep. These activities together with physiological stimuli administered to the host may cause brief disturbances of the periodicity cycle of the microfilariae but recovery following cessation of such stimuli or activity is rapid (see reviews by Hawking, 1965, 1967).

Several explanations were put forward to account for the production of filarial periodicity. From these, two principal hypotheses have emerged which

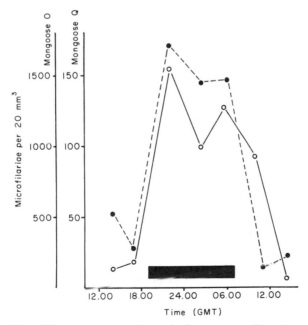

Figure 1. Periodicity of *Dipetalonema setariosum* in the mongoose *Herpestes sanguineus.* Solid black bar represents period of darkness. ●, Mongoose O; ○, mongoose Q.

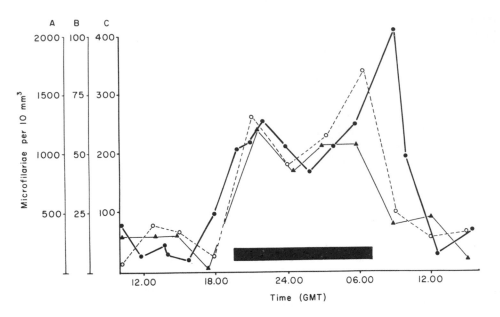

Figure 2. Periodicity of *Dipetalonema setariosum* in the mongoose. Curve A obtained from a natural infection. Curves B and C obtained six and ten months after reinfection three years after the date of Curve A. ●, A; ▲, B; ○, C.

differ with regard to the stage of the life cycle responsible for periodicity. Myers, in 1881, proposed that the rhythm was produced by the adult worm acting in concert with the host and the microfilariae which were produced circulated in the blood for a brief period of 12 h and were then destroyed by the host to be replaced by a new brood, i.e. they were short-lived. This concept was later modified by Lane (1929), who suggested that synchronicity was achieved by the simultaneous parturition of all of the female worms present in a host. Manson in 1882 attributed the production of the rhythm to the microfilariae themselves and suggested that parturition was continuous and that the microfilariae migrated to and from the peripheral blood each night and were therefore (by implication) long-lived (see reviews of Lane, 1948; Hawking & Thurston, 1951). Much of the research during the ensuing 50 years was carried out in the light of these hypotheses and particular emphasis was placed upon determining the longevity of the microfilariae, principally by means of experiments in which large numbers of microfilariae were transfused from one host to another. The results have been variable. In almost all cases there was rapid disappearance of the majority of the transfused microfilariae but of those which persisted, some survived for periods in excess of 24 h and in some species for periods of several years. Hawking (1953) transfused microfilariae of *Dirofilaria repens* into an uninfected dog and showed that the transfused microfilariae could not only survive for a considerable period but could maintain their periodic cycle in the absence of the adult worm. Hawking *et al.* (1963) transfused microfilariae of *Edesonfilaria malayensis* from a monkey whose activity cycle had been reversed, to a laboratory bred recipient monkey on a normal activity cycle, i.e. in opposite phase. In this experiment, the transfused microfilariae remained on the cycle of the donor monkey for a period of approximately 8 h before adapting to that of the recipient animal, indicating that the microfilariae themselves possessed a weak rhythm which quickly became entrained by that of the host. These and other similar observations support Manson's hypothesis that periodicity is due to the migration of the microfilariae.

Largely on evidence from histological studies of tissues obtained at post-mortem examination, several workers have reported the presence of large numbers of microfilariae in the lungs and it was suggested by Manson in 1898 that this was the location of the microfilariae when they were absent from the peripheral blood. By taking biopsies, at different times during the periodic cycle, from monkeys and dogs infected with *Dirofilaria* spp., Hawking & Thurston (1951) established that the lungs constituted a reservoir of microfilariae at all times but less microfilariae were present there during the period when a considerable number was present in the peripheral blood. Attempts by these workers to observe microfilariae directly in living animals were not successful. Kawaski (1958) and Shibata (1965) employed venous and cardiac catheterization in humans infected with *W. bancrofti* and dogs infected with *D. immitis* and found that the distribution of microfilariae in the living subject accords with that reported by indirect means and further that the appearance of microfilariae in the peripheral blood is due to their liberation from the lung reservoir.

The accumulation of microfilariae in the lungs appears to be due to an active response on the part of the parasites themselves but the means by which they are maintained in the lungs is not clear. Theories which postulate that parts of the lungs close down during the day are untenable, for in some areas of the world, microfilariae of similar size and morphology but with differing periodicities exist side by side in the same species of host, as for example the two strains of *W. bancrofti* in man. It has been proposed also that the microfilariae may adhere to the walls of the vessels of the lungs or attach themselves by the anterior hook but no such mechanisms were observed by Hawking & Clark (1967), who studied the microfilariae of *Dipetalonema witei* in the lungs of birds by cinematography. The microfilariae were continuously active and these authors suggested that they hold themselves within the lungs by continuous migration within its network of vessels rather than by remaining fixed at one point.

Accumulation of microfilariae in the lungs of the host occurs in some species of filariae which do not show periodicity (Wenk & Hofler, 1967; Worms *et al.*, 1961). Further, in those species which are periodic, it has been shown that the numbers of microfilariae present in the host are much greater than would be estimated by calculations based on the number present in the peripheral blood at the peak of the periodic cycle (Pacheco & Orihel, 1968; Greenough & Buckner, 1969).

It therefore appears that only a portion of the microfilarial population in a host takes part in the periodic cycle. We have no knowledge of the size of this portion nor do we know how frequently an individual microfilaria takes part in this cycle nor the number of cycles it undergoes. Undoubtedly microfilariae are capable of survival for considerable periods in hosts with no previous experience of filarial infection but the length of life of microfilariae in infected animals is not known. Wong (1964a) transfused microfilariae of D. immitis into dogs in which a patent infection was present; in these animals an increased microfilaraemia was observed for several days, suggesting that the transfused microfilariae had survived for this period. In artificially immunized dogs, Wong (1964b) found that microfilariae disappeared rapidly and it would appear that although microfilariae are capable of survival and periodic migration for a considerable period of time, their opportunity to do so in an infected individual may be severely curtailed by the immune response of the host. The population dynamics of microfilariae clearly offer a field for more research.

The maintenance and control of the periodic cycle, i.e. those factors which induce microfilariae to accumulate in or to leave the reservoir in the lungs has attracted considerable attention. Mention has been made already of the ability of the parasite to adjust its cycle, after a brief delay, to changes in the host's activity cycle. Such observations suggested that the maintenance and control of the cycle is dependent upon a factor or factors among the many circadian rhythms of the host. Considerable research has been devoted to their study and this has been reviewed recently by Hawking (1967). It has been established that the periodicity of microfilariae has many of the criteria of a true circadian rhythm and microfilariae apparently possess a "biological clock", although its location and nature remain unknown. Microfilariae of different genera, species and strains differ in their response to various stimuli administered to the host, particularly to inspired gases (O_2, CO_2) or to injection of substances which affect the para-sympathetic nervous system (Hawking et al., 1964); some species are sensitive to changes in the body temperature of the host (Hawking et al., 1965, 1967). It appears therefore that different species of microfilariae achieve their periodicity in different ways. Hawking (1967) has formulated a general hypothesis concerning filarial periodicity in which he has proposed that microfilariae accumulate in the lungs of the host because conditions there, particularly with regard to oxygen tension, are favourable and he has further proposed that those species exhibiting periodicity may be placed into groups depending on the manner in which the response to oxygen tension is effected.

These proposals suppose the existence of sensory structures in microfilariae. The microfilaria has been considered for many years to be a modified embryo or pre-larva. Laurence & Simpson (1971) have re-examined the status of this stage in the filarial life cycle and from a study of the histochemistry and ultrastructure they have concluded that the microfilaria is a modified true first stage larva adapted for life in the capillaries of the vertebrate host. Ultrastructural studies by Kozek (1968) and McLaren (1969) have revealed the presence of paired ciliary structures at the anterior and posterior extremities of the microfilaria for which a sensory function has been suggested.

24-h rhythms in trypanosomes

A 24-h periodic cycle, which like that of microfilariae appears to be due to a migration of individuals has been found in the Trypanosoma rotatorium complex in frogs. Southworth et al. (1968) found that in Rana clamitans the trypanosomes migrated to the peripheral circulation from a reservoir in the kidney, the peak numbers being present in the blood at about midday. The periodic cycle was altered by variation of photoperiod and these authors suggested that the cycle was related to light intensity, although reception of this stimulus was not via the visual pathway or the pineal gland complex of the host (Southworth *et al.,* 1968; Mason, 1970). Seed (1970) has also studied the *T. rotatorium* complex in *Rana clamitans* and has found that two morphological forms within the complex exhibit periodicities of opposite phase. Bardsley & Harmsen (1969, 1970) studied the behaviour of the *T. rotatorium* complex in *Rana catesbiana.* In this host, the peripheral parasitaemia was markedly affected by temperature and in frogs maintained at a constant temperature of 10°C a diurnal periodic cycle was not apparent. At this temperature, an increased parasitaemia was obtained by excitation of the host or by administration of adrenaline and these authors have suggested that the cycle is linked with the level of metabolic activity of the host.

Periodicity in trypanosomes of other vertebrate hosts has not been found but there are a few observations which suggest that such cycles may exist. Hornby & Bailey (1931) reported a diurnal variation in the numbers of *T. congolense* in the peripheral blood of an ox. They found that the number of trypanosomes present in the blood of the ear was higher at 07.00 than at 14.00 h and that this relationship was inversely proportional to ambient temperature. This relationship was affected by alteration of photoperiod and was subject to transient disturbance by the administration of certain drugs. No complete 24-h observations were made, however. That periodicity may exist also among avian trypanosomes is suggested by the observation of Laveran & Mesnil (1912 : 824), "En été, les trypan arrivent à ne plus être rares dans la circulation périphérique surtout quand l'examen est fait de nuit."

The observations on trypanosomes from different classes of vertebrates although unsubstantiated by detailed figures do suggest that periodicity among the trypanosomes may be of widespread occurrence and worthy of further investigation.

Periodicity in the Haemosporina

In the *Plasmodiidae,* rhythms are manifested not by the migrations of individuals but by their reproduction (although a form of migration may be said to exist in those species in which schizogony is confined to the deep tissue capillaries). The rhythmic recurrence of fevers in human malaria has been known since ancient times and it was demonstrated in the late 19th century that these fevers were related to the synchronous division of the asexual forms of the parasite. The similarity of these rhythms to the periodicity of microfilariae was noted and it was found that the timing of the division could be affected by alteration of the activity pattern of the host, and that it could be disturbed for relatively brief periods by administration of various stimuli to the host (Stauber, 1939). It was established also that different species and strains of *Plasmodium* exhibit rhythms of different phases and degrees of synchronicity. These rhythms, however, differ from microfilarial periodicity in two major aspects. Firstly the rhythms are shown by the asexual stages of the life cycle which are not involved in transmission. (A periodic cycle of development in the gametocytes of *Plasmodium cathemerium* was demonstrated by Shah (1934) and Gambrell (1937), but these authors did not explore fully the relationship with the asexual cycles and their observations have been largely overlooked.) Secondly, many of the rhythms have a periodicity which is greater than 24 h. Hawking *et al.* (1968) emphasized, however, that the known periodicities are, with two exceptions (*Plasmodium gallinaceum* and *Plasmodium relictum*), simple multiples of 24 h and that the time of schizogony under natural conditions is a character constant for a particular strain or species of parasite. These workers studied in detail the cycles of development of the gametocytes of three highly synchronous species, *Plasmodium knowlesi* and *Plasmodium cynomolgi* in monkeys and *P. cathemerium* in birds which have periodicities of 24, 48 and 24 h respectively. In these species it was found that in batches of mosquitoes fed at four-hourly intervals throughout the course of one or two cycles of asexual reproduction, a distinct cycle of infectivity was present which was shown by the numbers of oocysts which developed; maximum numbers of oocysts developed when the mosquitoes were fed at night. The infectivity was correlated with a maturation of the gametocytes as shown by their readiness to exflagellate. Morphological development of the gametocytes was also demonstrated and the peak periods of infectivity and exflagellation were found to coincide with the presence in the blood of maximum numbers of morphologically mature gametocytes. It appeared, that contrary to the established view that gametocytes are long-lived, they were relatively short-lived and their period of ripeness was of brief duration. These authors suggested that, as gametocytes and asexual forms may arise from the same merozoites at schizogony, the synchronous cycles of asexual reproduction are linked with the production of these relatively short-lived gametocytes. It was also found that the body temperature cycle of the host plays a significant role in the control and timing of these cycles but the manner in which this control is effected and the stage in the parasite life cycle that is affected remains unknown.

Other members of the *Haemosporina* differ from the *Plasmodiidae* in that gametocyctes only are present in the peripheral blood. Hitherto no cycles of development or infectivity have been detected among them but experience with *Plasmodium* suggests that such cycles could perhaps be investigated by examination of the exflagellation of gametocytes. In this context it is of interest that Novy & Macneal (1905: 263) commenting on the presence of "Leucocytozoa" in their trypanosome-infected birds note that "the formation of microgametes *particularly during the morning hours* is very prompt and may attract attention where otherwise the parasite would be unnoticed" (my italics).

SEASONAL RHYTHMS

In many blood parasites, rhythms exist which are less well defined than 24-h periodicities and occur on an approximately annual basis. These rhythms have been correlated with a particular period in the annual cycle and have been termed seasonal periodicities.

Microfilariae

Kubo (1938) found that, although the nature of the 24-h periodic cycle of *D. immitis* in dogs in China remained unaltered, the numbers of microfilariae present at the peak of the cycle were greater during the summer months than during those of the winter. A similar seasonal elevation in microfilaraemia in *D. immitis* has been reported by Newton (1968) in dogs in North America and by Hawking (1967) in England. This seasonal periodicity also occurs in the closely related species of *D. repens* in dogs and is maintained when the host is transferred to conditions in which seasonal variation in day length and in ambient temperature is markedly reduced (Fig. 3). The mechanism underlying

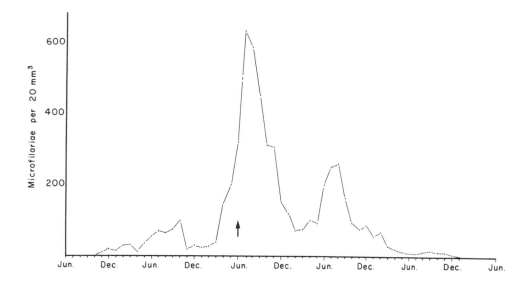

Figure 3. Seasonal periodicity of *Dirofilaria repens* in the dog. Curve of mean monthly microfilaraemia. Microfilaria counts taken at 16.30 hours. Arrow indicates point at which the dog was transferred from a kennel in which it was subject to natural seasonal variation in light dark schedule and temperature to one in which this variation was greatly reduced.

this phenomenon is not known. Katamine *et al.* (1970) found that by artificially increasing the temperature of the environment during the winter months, an increased level of microfilaraemia of *D. immitis* in dogs was obtained and these authors have suggested that this seasonal periodicity is related to the increase in temperature during the summer in temperate regions.

Protozoan blood parasites

Detailed long-term studies of parasitaemia in individual hosts with protozoal infection have been few, but increased levels of parasitaemia during the spring and summer months for the majority of protozoal blood parasites are commonly found in field surveys of populations of hosts of all classes of terrestrial vertebrates.

In those members of the *Haemosporina* in which only gametocytes occur in the blood, gametocyte production is dependent upon the activity of schizonts in the tissues and the appearance of gametocytes in the peripheral blood, following a period of absence or very low parasitaemia, may be due to a resumption of activity by these tissue forms.

In the *Plasmodiidae,* both gametocytes and asexual forms are present in the blood and the elevated parasitaemias seen in the spring and summer months may owe their origin to an increased activity of persistent blood forms, to a true relapse of exo-erythrocytic forms or, as in *P. vivax,* to the existence of strains with an extended prepatent period enabling infections acquired in the summer of one year to become patent the following year (Garnham, 1967).

The mechanisms underlying these phenomena are poorly understood. In birds, it has been shown that a relapse may be provoked earlier than usual by a variety of factors including prolongation of day length (Chernin, 1952) administration of cortisone (Applegate, 1970) or administration of sex hormones (Haberkorn, 1968; Rogge, 1965) and it would appear that relapses may be associated with the physiological changes coincident with the onset of the host's reproductive activity in the spring. Among mammals, a coincidence of reproductive biology of the host and elevated parasitaemia has been reported for *Polychromophilus melanipherus* in bats in Australia by Dew & McMillan (1970) and it is of interest that both the maximum occurrence of this parasite and the reproductive period of the host in the Southern Hemisphere is the reverse of that in the northern, indicating a definite seasonal correlation.

SIGNIFICANCE OF RHYTHMS IN BLOOD PARASITES

From the foregoing survey it is apparent that for certain groups of blood parasites, rhythmic behaviour is well established and the few observations that we have regarding other groups suggest that the phenomenon may be of widespread occurrence throughout the blood parasites as a whole. It is apparent also, that these rhythms are produced and controlled in different ways by different species or groups of parasites. There has been much speculation upon the significance of these rhythms in the life histories of the parasites. Writing generally of biological rhythms Cloudsley Thompson (1961) states, "Unless cyclical fluctuations in animal populations have an ultimate biological significance it is difficult to understand how the phenomenon can persist in nature."

It has long been held that microfilarial periodicity is a matter of intimate association between the parasite and the feeding habits of the insect host. Thus the nocturnally periodic form of *W. bancrofti* is transmitted by the night-biting *Culex fatigans,* the diurnally subperiodic form by the day-biting *Aedes polynesiensis* and there is similar correlation between the periodicity cycles of the microfilariae and the biting cycles of the vectors of the two forms of *Loa loa* and *Brugia malayi.* This relationship is not confined to the microfilariae; there is a correlation between the period at which the transmission forms of some blood protozoan parasites are at maximum concentration in the peripheral blood and the period at which the arthropod hosts are either actively feeding or are able, by virtue of a cessation of host activity, to do so. Thus Hawking *et al.* (1968) have suggested that the timing of the cycles of the malaria parasites studied by them was such as to ensure that the short-lived gametocytes were present in the blood at night, when potential vectors were feeding. Bardsley & Harmsen (1969) have suggested that, in the case of *Trypanosoma rotatorium* in the bull frog, the elevated parasitaemias, observed at a time when the frogs were basking, would coincide with a period of maximum vector-host contact, if the vector was a terrestrial one. Similarly Seed (1970) has suggested that the periodicities of opposite phase, in two members of the *T. rotatorium* complex, placed one form at maximum numbers during the aquatic phase of the host's daily activity pattern and hence would facilitate transmission by an aquatic vector (a leech) and the other form at peak numbers during the terrestrial phase, when the vector is presumed to be a mosquito. These suggestions, however, highlight the difficulty of discussion of such relationships for, in a great many instances, there is no knowledge of the vectors of the blood parasites or of their activity patterns in nature. In addition to the circadian cycles, the seasonal cycles, particularly in temperate regions, may also be correlated broadly with those periods in which arthropod vectors are most abundant or with a period when hosts are congregated together for reproductive purposes as in the case of *Polychromophilus* infection in bats.

Clearly this coincidence of maximum abundance of parasite and vector is a very favourable adaptation which ensures that the chances of transmission are maximal. However why does the parasite not achieve this by the continuous presence of transmission forms in the peripheral blood? Hawking & Thurston (1951) have suggested that the lungs may be a particularly attractive site for microfilariae and among other groups of blood parasites "deep tissue reservoirs" are known or thought to exist. The parasites require to be in the peripheral blood for transmission and "periodicity" is an adaptation which ensures that this occurs. It has generally been held that presence of the parasites in the peripheral blood is a passive phase in the daily cycle and that accumulation in a reservoir such as the lungs represents an active phase in response to an attraction. It is possible however, that this accumulation is an avoidance reaction and that the peripheral blood may in some way be unfavourable to them. This concept has been little developed with regard to periodicity.

The accumulation of microfilariae in the lungs both in periodic and nonperiodic species however, suggests that a fundamental character of microfilariae is to spend at least part of each 24 h in the tissues and part in the general circulation. The apparent lack of a cycle in a nonperiodic species may be due to the microfilariae acting out of phase with one another; a cycle would only become detectable by individuals acting in unison. The well defined periodicities of some species may have arisen by selection pressure applied by a vector with a relatively brief period of feeding activity. This would tend to select a number of individuals from an asynchronous population which are similar in phase. Continued selection could lead to the development of a host-parasite-vector ecosystem in which the cycles of the components were synchronized and stable. The stability in various hosts of the phase of the cycle in periodic microfilariae contrasted with subperiodic forms has already been discussed. Duke (1964) has shown that hybridization of the two periodic forms of *L. loa* produced forms with intermediate periodicities not inconsistent with the inheritance of the phase of the periodicity along simple Mendelian lines. It is possible therefore to consider the nonperiodic state as representing a genetic pool of mixed periodicities from which periodic strains may be selected. In the periodicity complexes of *L. loa, B. malayi* and *D. reconditum* such selection is now in process, and the parasite is still susceptible to influences both of host and vector. Among the blood protozoa, a similar situation may exist in the *P. relictum* complex. This species has a wide host range both in vertebrate and insect hosts, it is poorly synchronous and has a periodicity determined as 30-36 h. Corradetti & Neri (quoted in Garnham 1966, p. 553) have suggested that from this species has arisen the species *Plasmodium sub-praecox* which is restricted to owls and has a periodicity of 24 h.

Clearly such hypotheses are very tenuous in the present state of our knowledge but should genetically "fixed" strains and species of parasites have arisen in this manner, their continued existence would depend on the ability of the parasite to maintain the timing and synchronicity of its periodicity, in order to maintain its relationship with its vector. From this point of view rhythmic behaviour assumes an undoubted significance in relation to parasite transmission.

REFERENCES

ALCOCK, A. & MANSON BAHR, P., 1927. The life and work of Sir Patrick Manson. London: Cassell.
APPLEGATE, J. E., 1970. Population changes in latent avian malaria infections associated with season and corticosterone treatment. *J. Parasit., 56*(3): 439-443.
BARDSLEY, J. E. & HARMSEN, R., 1969. The trypanosomes of Ranidae I. The effects of temperature and diurnal periodicity on the peripheral parasitaemia in the bullfrog (*Rana catesbeiana Shaw*). Can. J. Zool., 47(3): 283-288.
BARDSLEY, J. E. & HARMSEN, R., 1970. The trypanosomes of Ranidae II. The effects of excitation and adrenalin on the peripheral parasitaemia in the bullfrog (*Rana catesbeiana Shaw*). Can. J. Zool., 48(6): 1317-1319.
BOUGHTON, D. C., BYRD, E. E. & LUND, H. O., 1938. Microfilarial periodicity in the crow. J. Parasit., 24: 161-165.
CHERNIN, E., 1952. The relapse phenomenon in the *Leucocytozoon simondi* infection of the domestic duck. *Am. J. Hyg., 56*(2): 101-118.
CLOUDSLEY-THOMPSON, J. L., 1961. *Rhythmic activity in animal physiology and behaviour.* New York: Academic Press.
DEW, B. B. & McMILLAN, B., 1970. Seasonal variation of *Polychromophilus melanipherus* (*Sporozoa: Haemoproteidae*) in the bent winged bat *Miniopterus schreibersi* (*Chiroptera*) in New South Wales. *Parasitology, 61*: 161-166.
DUKE, B. O. L., 1964. Studies on loiasis in monkeys IV. Experimental hybridisation of the human and simian strains of *Loa. Ann. trop. Med. Parasit., 58*(4): 390-408.
DUKE, B. O. L. & WIJERS, D. J. B., 1958. Studies on loiasis in monkeys I. The relationship between human and simian *Loa* in the rain forest zone of the British Cameroons. *Ann. trop. Med. Parasit., 52*: 158-175.
EDESON, J. F. B. & WILSON, T., 1964. The epidemiology of filariasis due to *Wuchereria bancrofti* and *Brugia malayi. A. Rev. Ent., 9:* 245-268.
GAMBRELL, W. E., 1937. Variations in gametocyte production in avian malaria. *Am. J. trop. Med., 17:* 689-729.
GARNHAM, P. C. C., 1966. *Malaria parasites and other haemosporidia.* Oxford: Blackwell Scientific Publications.
GARNHAM, P. C. C., 1967. Relapses and latency in malaria. *Protozoology, 2:* 55-64.
GREENOUGH, W. B., III & BUCKNER, D., 1969. Removal of microfilariae from unanaesthetised dogs by continuous flow centrifugation. *Trans. R. Soc. trop. Med. Hyg., 63*(2): 259-262.
GUBLER, D. J., 1966. A comparative study on the distribution incidence and periodicity of the canine filarial worms *Dirofilaria immitis* Leidy and *Dipetalonema reconditum* Grassi in Hawaii. *J. Med. Ent., 3*(2): 159-167.
HABERKORN, A., 1968. Zur hormonellen Beeinflussung von *Haemoproteus*-Infektion. *Z. Parasitenk., 31:* 108-112.
HAWKING, F., 1953. The periodicity of microfilariae III. Transfusion of microfilariae into a clean host. *Trans. R. Soc. trop. Med. Hyg., 47*(1): 82-83.
HAWKING, F., 1962. Microfilaria infestation as an instance of periodic phenomena seen in host parasite relationships. *Ann. N.Y. Acad. Sci., 98:* Art 4, 940-953.

HAWKING, F., 1965. Advances in filariasis especially concerning periodicity of microfilariae. *Trans. R. Soc. trop. Med. Hyg.*, *59*(1): 9-25.

HAWKING, F., 1967. The 24 hour periodicity of microfilariae: biological mechanisms responsible for its production and control. *Proc. R. Soc. (B)*, *169:* 59-76.

HAWKING, F. & CLARK, J. B., 1967. The periodicity of microfilariae XIII. Movements of *Dipetalonema witei* microfilariae in the lungs. *Trans. R. Soc. trop. Med. Hyg.*, *61*(6): 817-826.

HAWKING, F. & THURSTON, J. P., 1951. The periodicity of microfilariae I. The distribution of microfilariae in the body. II. The explanation of its production. *Trans. R. Soc. trop. Med. Hyg.*, *45*(3): 307-340.

HAWKING, F., ADAMS, W. E. & WORMS, M. J., 1964. The periodicity of microfilariae VII. The effect of parasympathetic stimulants upon the distribution of microfilariae. *Trans. R. Soc. trop. Med. Hyg.*, *58*(2): 178-194.

HAWKING, F., GAMMAGE, K. & WORMS, M. J., 1965. The periodicity of microfilariae X. The relation between the circadian temperature cycle of monkeys and the microfilarial cycle. *Trans. R. Soc. trop. Med. Hyg.*, *59*(6): 675-680.

HAWKING, F., WORMS, M. J. & GAMMAGE, K., 1968. 24- and 48-hour cycles of malaria parasites in the blood; their purpose, production and control. *Trans. R. Soc. trop. Med. Hyg.*, *62*(6): 731-760.

HAWKING, F., WORMS, M. J. & WALKER, P. J., 1963. The periodicity of microfilariae IX. The transfusion of microfilariae (*Edesonfilaria*) into monkeys at a different phase of the circadian rhythm. *Trans. R. Soc. trop. Med. Hyg.*, *59*(1): 26-41.

HAWKING, F., MOORE, P., GAMMAGE, K. & WORMS, M. J., 1967. Periodicity of microfilariae XII. The effect of variations in host body temperature on the cycle of *Loa loa, Monnigofilaria setariosa, Dirofilaria immitis* and other filariae. *Trans. R. Soc. trop. Med. Hyg.*, *61*(5): 674-683.

HORNBY, H. E. & BAILEY, H. W., 1931. Diurnal variation in the concentration of *Trypanosoma congolense* in the blood vessels of the ox's ear. *Trans. R. Soc. trop. Med. Hyg.*, *24*(5): 557-564.

KATAMINE, D., AOKI, Y. & IWAMOTO, I., 1970. Analysis of microfilarial rhythm. *J. Parasit.*, *56*(4 sect. 2 pt. 1): 181.

KAWASKI, K., 1958. Pathophysiological studies on filariasis (F. 10). An approach to the mechanism of microfilarial periodicity by means of venous catheterisation. *Med. J. Kagoshiwa Univ.*, *9*(6): 34-60. (In Japanese, English Summary.)

KOZEK, W. J., 1968. Unusual cilia in the microfilaria of *Dirofilaria immitis. J. Parasit.*, *54:* 838-844.

KUBO, M., 1938. The daily and seasonal periodicity of *Microfilaria immitis* in the peripheral blood of the dog. *China med. J.*, *2*(Suppl.): 375-384.

LAING, A. B. G., 1961. Influence of the animal host on the microfilarial periodicity of *Brugia malayi. Trans. R. Soc. trop. Med. Hyg.*, *55:* 558.

LANE, C., 1929. The mechanism of filarial periodicity. *Lancet, 1:* 1291-1293.

LANE, C., 1948. Bancroftian filariasis. Biological mechanisms that underlie its periodicity and other of its clinical manifestations. *Trans. R. Soc. trop. Med. Hyg.*, *41*(6): 717-784.

LAURENCE, B. R. & SIMPSON, M. G., 1971. The microfilaria of *Brugia*: a first stage nematode larva. *J. Helminth.*, *45*(1): 23-40.

LAVERAN, A. & MESNIL, F., 1912. *Trypanosomes et Trypanosomiasis.* Paris: Masson et Cie.

LOW, G. C. & MANSON-BAHR, P. H., 1933. Some recent observations on filarial periodicity with a clinical and laboratory report by A. H. Walters. *Lancet, 1:* 466-468.

MACKENZIE, S., 1882. A case of filarial haemato-chyluria. *Trans. path. Soc. Lond.*, *33:* 394-410.

MANSON, P., 1882. *Filaria sanguinis hominis* and fever. *Lancet, 1:* 289-290.

MANSON, P., 1898. *Tropical diseases*, 1st ed. London: Cassell.

MARINKELLE, C. J., 1970. Observaciones sobre la periodicidad de las microfilarias de *Ochoterenella* en *Bufo marinus* de Colombia. *Rev. Biol. Trop.*, *16*(2): 145-152.

MASON, G., 1970. The diurnal rhythm of *Trypanosoma rotatorium* in *Rana clamitans*. Investigation of photoreceptors and physiological control. *J. Parasit.*, *56* (4 sect. 2. No. 1): 228.

McLAREN, D. J., 1969. Ciliary structures in the microfilaria of *Loa loa. Trans. R. Soc. trop. Med. Hyg.*, *63*(2): 290-291.

MYERS, W. W., 1881. Observations on *Filaria sanguinis hominis* in South Formosa. *Med. Rep.*, *21:* China Customs Gazette.

NEWTON, W. L., 1968. Longevity of an experimental infection with *Dirofilaria immitis* in a dog. *J. Parasit.*, *54*(1): 187-188.

NEWTON, W. L. & WRIGHT, W. H., 1956. The occurrence of a dog filariid other than *Dirofilaria immitis* in the United States. *J. Parasitol.*, *42:* 246-258.

NOVY, F. G. & MACNEAL, W. J., 1905. On the trypanosomes of birds. *J. infect. Dis.*, *2:* 256-308.

PACHECO, G. & ORIHEL, T. C., 1968. Relationship between detectable microfilaraemia and total population of microfilariae in monkeys infected with *Dirofilaria corynodes. J. Parasit.*, *54*(6): 1234-1235.

PENNINGTON, N. E. & PHELPS, C. A., 1969. Canine filariasis on Okinawa, Ryuku Islands. *J. med. Ent.*, *6*(1): 59-67.

ROGGE, D., 1965. *Haemoproteus* and other blood protozoa in Central European Birds. *Prog. Protozool. (1965) Excerpta Medica Foundation Int. Congr.*, Ser. No. 91: 176-177.

SEED, J. R., 1970. Diurnal and seasonal rhythms in parasitaemia levels of some trypanosomes infecting *Rana clamitans* from Louisiana. *J. Parasit.*, 56 (4 sect. 2 pt. 1): 311.

SHAH, K. S., 1934. The periodic development of sexual forms of *Plasmodium cathemerium* in the peripheral circulation of canaries. *Am. J. Hyg.*, *19:* 392-403.

SHIBATA, S., 1965. Experimental studies on the periodicity of microfilariae VI. Antemortem studies of distribution of microfilariae in the body. *Endem. Dis. Bull. Nagasaki Univ.*, 7(1): 1-11.

SOUTHWORTH, G. C., MASON, G. & SEED, J. R., 1968. Studies in frog trypanosomiasis. I. A 24-hour cycle in the parasitaemia level of *Trypanosoma rotatorium* in *Rana clamitans* from Louisiana. *J. Parasit.*, *54:* 255-258.

STAUBER, L., 1939. Factors influencing the asexual periodicity of avian malarias. *J. Parasit.*, *25:* 95-116.

WENK, P. & HOFLER, W., 1967. Einfluss von Sauerstoffmangel in der Atemluft auf der Mikrofilarämie (*Litomosoides carinii*) (*Filariidae*) bei der Baumwollratte *Z. Tropenmed Parasit.*, *18*(4): 396-402.

WHO Expert Committee on Filariasis 2nd Report, 1967. *Tech. Rep. Ser. Wld Hlth Org.:* 359.

WONG, M. M., 1964a. Studies on microfilaremia in dogs. I. A search for the mechanisms that stabilise the level of microfilaremia. *Am. J. trop. Med. Hyg.*, *13*(1): 57-65.

WONG, M. M., 1964b. Studies on microfilaremia in dogs. II. Levels of microfilaremia in relation to immunologic responses of the host. *Am. J. trop. Med. Hyg.*, *13*(1): 66-77.

WORMS, M. J., TERRY, R. J. & TERRY, A., 1961. *Dipetalonema witei*, filarial parasite of the jird *Meriones libycus* I. Maintenance in the laboratory. *J. Parasit.*, *47*(6): 963-970.

DISCUSSION

N. A. Croll

Have any investigations been made on the behaviour of microfilariae *in vitro?*

M. Worms

Very few *in vitro* investigations appear to have been made. This is probably due principally to the technical difficulties involved particularly in effecting adequate controls.

F. E. G. Cox

The incidence of *Haemoproteus* and *Leucocytozoon* in West African birds does not seem to be seasonal. In most temperate regions, where dawn and dusk occur at different times in relation to the "working day" during the course of the year, it is important to separate the effects of possible seasonal variations from variations in the actual times at which blood smears are made. Daily changes in parasitaemia could easily be interpreted as seasonal ones.

M. Worms

I agree that an awareness of the possible influence of rhythmic behaviour upon observations made during the "working day" is desirable. The recording of the time at which observations are made might lead to the discovery of periodicities in those groups in which we have only suspected their existence. Also in experiments in which serial observations are to be made over an extended period, these should be made at approximately the same hour of day and the time of year noted. Clearly observations upon microfilariaemia made upon for example *Dirofilaria repens* from January to June would differ from those made in July to December.

D. S. Bertram

One of the plasmodial infections with a periodicity other than 24 h, or a multiple of 24 h, is *P. gallinaceum* of fowls with a 36-h cycle. This must imply a strong endogenous rhythm not readily influenced by cues from the 24-h circadian rhythm of the fowl host. It also means that this particular parasite is cleverly adapted to its mosquito transmission, alternating (in the individual bird) a night periodicity on one occasion with a diurnal periodicity in succession 36 h later, and so on. It backs both options for a vector, or vectors.

M. Worms

We have studied recently the periodicity of *Plasmodium gallinaceum* and by manipulating the frequency and timing of sub-inoculation we have obtained a moderately synchronous strain. The schizogonic cycle has remained at 32-36 h. As judged by their readiness to exflagellate, the gametocytes appear to remain mature for a longer period (approximately 17 h) than those of the plasmodia with 24- or 48-h cycles. It may be that this relatively long period of maturity which would cover a number of activity periods of potential vectors has overcome any tendency for selection by a particular vector, and has not therefore necessitated any strong dependence upon cues from the circadian rhythms of the host.

Some aspects of mosquito behaviour in relation to the transmission of parasites

M. T. GILLIES

School of Biological Sciences, University of Sussex, Brighton

Host choice is the most important characteristic influencing the role of mosquitoes as vectors of parasites. Genera such as *Aedes* are highly plastic in feeding behaviour, and choice of host is mainly determined by ecological factors and host availability. In some other groups of mosquitos clear-cut preferences for particular groups of animals exist, within which a hierarchy of preferred hosts can be recognized. Odour is the factor most likely to be involved in long range host choice. Flight pattern is seen as another behavioural factor influencing host/vector encounters. Certain sharply defined epidemiological situations can be recognized, determined by the mobility of the host and the ambit of the mosquito.

CONTENTS

HOST SELECTION IN MOSQUITOES

In viewing the epidemiology of vector borne diseases in terms of the behavioural patterns of vectors, host selection is clearly the single most important factor but despite its importance, it is still a very little understood subject. Laboratory workers on mosquitoes have sometimes tended to underestimate host-specificity in feeding habits, especially if they have worked with *Aedes aegypti* which can be persuaded to feed on anything from cows to caterpillars (e.g. Tempelis, Hayes, Hess & Reeves, 1970; Harris Riordan & Cooke, 1969). In contrast malaria workers, with their interests focussed on the human host, frequently employ the term "anthropophilic" for species of *Anopheles* caught habitually on man, with the implication that man is preferred above all other hosts. Similarly, the uncritical interpretation of the results of precipitin tests on the stomach contents of resting mosquitoes can be misleading.

One can recognize three components in the process of bringing the vector into contact with the host:

(i) innate host-specific tendencies of the vector;

  (ii) the habits and ecology of the host;

  (iii) the flight pattern and ecology of the vector.

Because of the difficulty of disentangling the first factor from the others, very little is known of the finer degrees of discrimination by the vector between one type of host and another, even though field observations are sometimes strongly suggestive of such behaviour. The difficulty arises from the fact that, although some species will attack a wide range of hosts when these are presented separately, they may show a tendency to concentrate on one type of host when a number of others are equally available. In other words, within the range of acceptable hosts a hierarchy exists, based on the relative attractiveness of different hosts (Flemings, 1958). A system of rating of this sort, however, has very little meaning, even if established under experimental conditions, unless one takes into account differences in size of host. As the results of Dow, Reeves & Bellamy (1957) and Downe (1960) showed, one must assume that both host-specific factors (odour) and non-specific or group factors (CO_2, convection currents) contribute to host selection. Consequently, in order to determine the existence of specific factors one must present the insects with alternative hosts of comparable size or surface area. If there is any gross discrepancy this is obviously not practicable. Another difficulty is that even quite small differences in the manner in which alternative hosts are presented can alter their relative attractiveness to blood-sucking insects.

  As a result, the only practical approach is to compare the responses of different vectors simultaneously to pairs or sets of alternative hosts. In this way, several species of insects are exposed to the same variation in output of non-specific factors by the hosts and any differences in behaviour can be attributed to inherent host-specific tendencies.

  An example of this may be provided by studies on the *Anopheles gambiae* complex in East Africa (Gillies, 1967). Two species were investigated, *A. gambiae* sp. A, a freshwater species, which, on the evidence of precipitin tests, feeds largely on man, and is a very important vector of malaria, and *Anopheles merus,* salt-water breeding, which, on the evidence of precipitin tests feeds largely on cattle (Iyengar, 1962), and is at most a very minor vector. In the East African coastal belt cattle are largely absent, whereas on Pemba Island, where *A. merus* is particularly abundant, there are large herds of cattle, which are kept outdoors at night some distance from villages and in close proximity to the salt marshes. In view of these circumstances it might be thought that the difference in feeding habits was solely due to differences in the availability of the two hosts. To test this a volunteer slept either in the same room as a calf or else in a partially screened compartment of the same room. Unfed females of the two species were released in the room over-night and fed specimens captured next morning. Precipitin tests were done to determine the source of blood: with only one host present both species fed readily on either man or the calf but with both hosts present the proportion of both species of mosquitoes feeding on the calf varied according to the experimental set-up. In all experiments we found that there were always about twice as many of the saltwater species *A. merus* feeding on the calf as there were of the freshwater *A. gambiae.* Thus we were able to show that the difference in behaviour observed in nature was only partly determined by host-availability, and that an innate preference for the human host was a characteristic of freshwater *A. gambiae.*

A slightly different approach is to take a limited number of hosts and to grade a variety of species of mosquitoes according to the proportions attracted in field trials to the different hosts. Reid (1961) did this in Malaya for the responses of 26 species to human and calf baits exposed in trap-nets well separated from each other in the field. The man : calf attraction ratio in 11 species of *Anopheles* ranged from 3.4 : 1 in the case of dark-winged *Anopheles barbirostris,* an important vector of malaria, to 1 : 82 in the case of *Anopheles vagus,* a non-vector. Thus, with more or less equal sized hosts, clear evidence was produced to show the existence of a range of relative preferences for one host over another.

At the group level there is reasonable evidence that certain species or species-groups of mosquitoes are restricted in their feeding habits to certain groups of vertebrates. For example, a critical examination of the relatively well studied *Anopheles* of tropical Africa (Gillies and De Meillon, 1968) shows that, of 72 species on which we have reasonable data, 7 feed habitually on man, 59 on any large mammal and 6 on small mammals.

A number of groups or species of largely ornithophilic mosquitoes are known, of which certain populations of *Culex pipiens pipiens* provide the best example. The dominance of these forms in north-west Europe contrasts with the hibernating man-biting *C. pipiens* of northern Russia (Morozov, 1965; Kupriyanova & Vorotnikova, 1967) and with the non-hibernating and mammal-biting populations of the Mediterranean area. In the case of the north-west European *C. pipiens* the selection of birds, and to a lesser extent of poikilotherms, seems to be absolute and mammals are never attacked. It is likely that a number of tropical species of *Culex* show the same behaviour.

More specialized still in their feeding preferences are a number of groups, such as the subgenera *Neoculex* (Crans, 1970), *Ficalbia* (*Mimomyia*) (Menon & Tampi, 1959) and certain species of *Uranotaenia* (Chapman, 1970). These species are mainly or exclusively restricted to feeding on poikilotherms, especially amphibia, and some of them probably play a role in the transmission of frog filarioids (Crans, 1969, 1970).

In contrast to these one may cite mosquitoes of the genus *Aedes,* most species of which show a very wide degree of host acceptance, some being known to feed on mammals, birds, reptiles, and amphibia (Hayes, 1961; Murphey, Burbutis & Bray, 1967). There is a species in Malaya that is known to attack mudskippers, *Periophthalmus* spp., as they emerge from the water and crawl over the muddy banks of mangrove swamps (Garcia & Jeffery, 1969). Species of this genus are the most universal of pest mosquitoes, and it is a fairly safe assumption that a large breeding population of almost any *Aedes* will constitute a threat to human economy or comfort.

This very brief survey of the different patterns of host selection shown by mosquitoes serves to illustrate the point that, while for some groups host selection is closely related to size and availability, in others feeding is restricted to certain groups of vertebrates within which a hierarchy of preferred hosts can be recognized. Instances in which man occupies a senior position in the hierarchy are known. From an epidemiological point of view, species with a narrow range of hosts are more likely to be of importance as vectors of parasitic diseases, such as malaria, while those that switch from one group of hosts to another according to local circumstances are more likely to be involved in arbovirus transmission. There are, of course, a number of exceptions to this, of which the transmission of the sub-periodic form of *Brugia malayi* is an obvious example (Wharton, 1962).

HOST DISCRIMINATION AT A DISTANCE

The next question to consider is how the vector discriminates between one host and another. Apart from surface factors, that are effective after the insect has made contact with the skin of the host, there are a number of influences operating at a distance. A useful way of looking at the problem is by the concept of the host-stream. By this one means the plume of host-conditioned air drifting down-wind from the host (Daykin, Kellogg & Wright, 1965; Gillies & Wilkes, 1969). It is this which provides a blood-sucking insect with long range information on the presence of a suitable host and, by responding to the chain of stimuli encountered as it approaches the bait, the mosquito is able to home onto the target with accuracy.

Taking the host stimuli individually, one can see that the close range factors—convection currents and, for day-biting species visual stimuli—could influence the choice of host, as suggested by Daykin (1967) for mosquitoes in vertical air currents. At all ranges odour is likely to provide the best information for enabling a vector to discriminate between species of hosts. In certain host/vector encounters, for example man or cattle and *Anopheles melas* in West Africa, odour has the added quality of enabling a vector to accept or ignore a particular host situated at a greater distance than others from which stimuli are received (Gillies & Wilkes, 1969, 1970).

On the other hand, carbon dioxide is a universal feature of all vertebrates, and at first sight its influence in host choice would appear to be slight. However, small hosts such as rodents and birds emit very small amounts of this substance. If bird-biting species were sensitive to carbon dioxide the large amounts given off by large animals could involve them in frequent diversionary flights towards a source of attractant, which might only be recognized as coming from an unacceptable host at relatively close range. Consequently it would be adaptive for such species to lose the capacity to respond to carbon dioxide, especially when encountered in the absence of the appropriate odours. Field evidence from West Africa on certain *Culex* is strongly suggestive of this sort of adaptation (Gillies & Wilkes, 1969, 1970).

Variation in these factors determines the choice of host by vectors, the chemically elusive substances contributing to host odour probably being the most important.

FLIGHT PATTERN AND HOST SEEKING

The second component involved in vector/host contact, namely the habits and ecology of the host, is not covered by the subject of the symposium. The third component, namely the flight pattern and ecology of the vector, is obviously interrelated with the second. For instance, the dispersal of the adults of a vector species will be influenced by the choice, and hence the location, of breeding sites and in turn will have a bearing on the chance of contact with one type of host rather than another. Of the behavioural aspects of parasite transmission it is specifically flight pattern and contact with the host-stream that I want to consider.

It is possible to recognize four main epidemiological situations:

(1) The *savannah/cultivated* situation, where breeding and feeding areas may be widely separated, and vector/host contact is the result of mosquitoes dispersing widely away from breeding areas.

(2) The *forest/woodland* situation, where mosquito activity is largely restricted to certain vegetation types and breeding is either within the vegetation or without. Vector/host contact is the result of man or other hosts penetrating into the forest.

(3) The *urban/domestic* nidus, where man and the mosquito share a common environment and contact is fairly close and continuous.

(4) The *deep cave* nidus, where the host and mosquito are in intimate contact at all times. This is an extreme example of the domestic nidus.

Table 1 shows the main characteristic of five examples of these host/vector situations.

Table 1. Epidemiological situations in relation to movements of vectors and hosts

Nidus	Savannah/ cultivated	Forest/ woodland	Urban/ domestic	Deep caves
Flight pattern of vector	Wide ranging ill defined ambit	Sharply defined ambit	Unrestricted short flights	Movement within single cavern
Movements of host	Static often indoors	Mobile	Mobile or static	Static within cave
Host attack released by:	Vector flying into host-stream	Host-stream impinging on resting or locally flying vector	Endogenous rhythm	Endogenous rhythm
Vector encounters with host stimuli	Sporadic	Sporadic	Frequent	Continuous
Examples	*Anopheles & malaria in Africa*	*Anopheles (Kerteszia) & human malaria in Neotropics*	*C. fatigans & Wuchereria bancrofti*	*Anopheles hamoni & bat malaria in Congo*

The savannah/cultivated situation

This situation is typical of the epidemiology of human malaria in many parts of the world. The breeding sites are often at a considerable distance (up to 2-3 km) from human settlements and the vectors have to disperse widely in their search for food (Fig. 1). These long range flights carry the penalty of

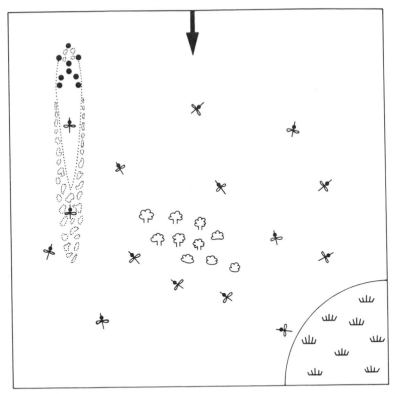

Figure 1. Diagram to illustrate the savannah/cultivated situation. Mosquitoes are shown dispersing across open country towards an isolated village. The black dots indicate the hosts, and the dotted lines depict the composite village host-stream. The large arrow indicates wind direction.

greater exposure to predators and the probability of a higher mortality per gonotrophic cycle. In East Africa, for instance, we found that the daily mortality of coastal populations of the *A. gambiae* complex, where breeding sites and feeding areas were closely associated, was 14.6%, whereas in an inland area, where breeding sites were mostly a kilometre or more from the villages, the daily mortality was 21% (Gillies & Wilkes, 1965). Differences in climate and species composition within the complex were not sufficient to account for this discrepancy and it seemed likely that the necessity for long range flights was at least partly responsible for the lower survival rates in the inland population and for the lower intensity of malaria transmission.

The mechanism of host finding in the savannah regions is not easy to understand. Host stimuli probably drift away from villages in dissected plumes extending for a considerable distance down wind, perhaps for a hundred metres or more. Contact with man will be the result of fortuitous contact with the village odour plume by mosquitoes dispersing from breeding sites. In West Africa in an area where the villages are compact and are 1.5 km or more from vast areas of mangrove, Giglioli (1965) found that *A. melas* tended to congregate near the villages and to fly fairly close to the interface between bush and cultivation. Our own work in the same district has shown that the great majority of mosquitoes fly in a generally up-wind direction when near the ground and there is some indication of down-wind flight at higher levels (Snow

& Wilkes, 1970). In other words, their direction of flight appears to be determined by the direction and perhaps the speed of the wind, not by the location of distant villages. Since (at certain seasons), in the district we worked in, the wind is at right angles to the line of direct flight between villages and the nearest breeding sites, mosquitoes must evidently reach the feeding areas by very roundabout flights.

The fact that the presence of a host is indicated by a plume of attractant-containing air parallel to the wind means that a wandering, partly up-wind and partly cross-wind, flight would give the best chance of interception by a hungry mosquito (Hocking & Khan, 1966). Successful interception would be enhanced on the one hand by larger aggregations of hosts, giving a longer host-stream, and on the other by the ability of vectors to detect host-specific odours at longer range than non-specific factors such as carbon dioxide. It is clear that sensory organs adapted to receive long range stimuli are of major importance to savannah-inhabiting species and that certain vectors, such as *A. melas,* appear to possess this attribute (Gillies and Wilkes, *loc. cit.*)

The forest/woodland situation

The basic characteristic of this situation is the low dispersal pattern of many mosquitoes, so that their ambit is limited to the area of more or less dense vegetation (Fig. 2). When man enters this habitat his host-stream drifts down

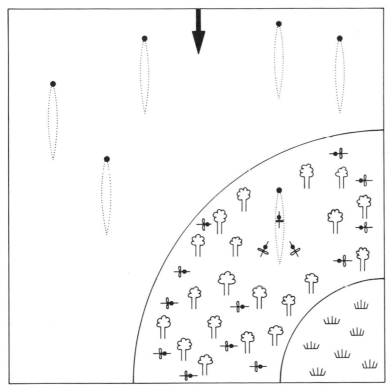

Figure 2. Diagram to illustrate the forest/woodland situation. Individual hosts are shown in the open country and in the forest, mosquitoes are restricted to the forest only. Legend as in Fig. 1.

wind on to resting or locally flying mosquitoes leading to the simultaneous activation of large numbers of insects. Their rate of arrival at the host follows a typical depletion curve with an initially high biting rate, falling off rapidly as all the mosquitoes within range of the host-stream are removed from the feeding population (Fig. 3). This sort of situation is well known to field workers who, if they are interested in making large collections, move their baits on from one site to another as the catching rate in each declines (Haddow, 1945).

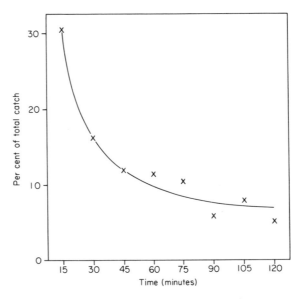

Figure 3. Number of *Aedes* attracted to a source of CO_2 in a Sussex wood in successive 15-minute periods. Catches made with a sticky trap baited with 1lb of dry ice (Gillies & Snow, 1970). Results expressed as geometric mean of six 2-hour daytime catches. Total catch, 1095 females of *A. cantans, A. annulipes, A. rusticus* and *A. punctor*.

The situation is perhaps commonest for species that are active in the daytime such as *Aedes* and *Psorophora*, and its importance in the transmission of arboviruses has long been recognized by yellow fever workers. But it is also important for *Anopheles* spp. of the subgenus *Kerteszia* which breed in bromeliads in tropical American forests and have been shown to be responsible for malaria transmission among people working in the forest by day (Charles, 1959). On the other hand, one must emphasize that this contact, important though it is in epidemiological terms, does not represent the normal feeding pattern of the vector which is more likely to be associated with some host resident in the forest. According to Forrattini, Lopes & Rabello (1968) it may well be birds in the case of *A. (Kerteszia) cruzii* in Brazil.

A situation similar to this concerns the rodents, harbouring *Plasmodium berghei*, that inhabit narrow belts of gallery forest in the Congo (Vincke, 1954). We know little about the behaviour of the vector, *Anopheles dureni millecampsi* except that its activity is apparently confined to these strips of vegetation. It seems likely that contact in this situation is the result of nocturnal encounters between foraging animals and locally dispersing mosquitoes.

The urban/domestic nidus

This focus exemplified by a section of a tropical town is shown diagrammatically in Fig. 4. Its essential features are the close intermingling of hosts, vectors and breeding sites throughout the area. The insects are therefore continually encountering attractant stimuli regardless of their physiological state. The frequency of effective man/vector contact is primarily determined by the activity cycles of the latter. Orientation to long range factors is not important. It is even possible that continual exposure leads to a degree of

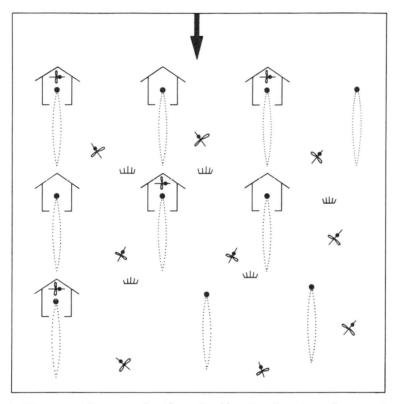

Figure 4. Diagram to illustrate urban/domestic nidus. Breeding sites adjacent to houses. Individual hosts and mosquitoes shown inside and outside houses throughout the area. Legend as in Fig. 1.

habituation to certain stimuli such as odour, so that only relatively high concentrations of host factors stimulate host-seeking responses. Transmission of a parasite is independent of the flight characteristics of the vector or the movements of the host, since wide ranging dispersal is rendered unnecessary by the frequency of arresting stimuli. The fact that urban *Culex fatigans* may nevertheless disperse quite extensively through a town (Lindquist, Ikeshoji, Grab, De Meillon & Khan, 1967) possibly reflects an intermittent receptivity of mosquitoes to host stimuli.

This situation is typical of the town dweller's encounters with *A. aegypti* by day or *C. fatigans* by night. In the past it has led to the great epidemics of yellow fever in the cities of tropical America; while at the present time

Bancroftian filariasis is increasing at an alarming rate as unbalanced development of tropical towns proceeds (Mattingly, 1969).

The deep cave nidus

An extreme example of association between host and vector is provided by cave mosquitoes such as *Anopheles hamoni,* which complete their entire life cycle in deep caves in the Congo basin (Adam, 1962, 1965). Their sole source of food is bats of the genus *Miniopterus,* with which they live in the closest proximity and to which they transmit a species of *Plasmodium* s. l. In the still air and total obscurity of these underground caverns, chemical stimuli must be of little help in locating the hosts, near to hand though they be. It seems likely, therefore, that they have to rely on convection currents in the close vicinity of the bats to achieve final contact with them.

CONCLUSIONS

From the examples I have quoted, it can be seen that in many cases host/vector associations are influenced by the flight pattern and orientation responses of mosquitoes as much as by the ecology and availability of the two hosts of the parasite. Attempts to isolate the specific chemical cues involved in host discrimination have not been very successful up to the present time, and it may be, as Hocking (1971) suggests, that blood-sucking insects recognize potential hosts for what they are by the totality of the chemical factors that we describe as odour.

REFERENCES

ADAM, J. P., 1962. Un Anophèle cavernicole nouveau de la République du Congo (Brazzaville): *Anopheles (Neomyzomyia) hamoni* n.sp. (Diptera-Culicidae). *Bull. Soc. Path. exot., 55:* 153-165.

ADAM, J. P., 1965. Les Culicidae cavernicoles du Congo et de l'Afrique intertropicale. *Annls Spéléol., 20:* 409-423.

CHAPMAN, H. C., 1970. Colonization of *Uranotaenia lowii* Theobald (Diptera:Culicidae). *Mosq. News, 30:* 262.

CHARLES, L. J., 1959. Observations on *Anopheles (Kerteszia) bellator* D. & K. in British Guiana. *Am. J. trop. Med. Hyg., 8:* 160-167.

CRANS, W. J., 1969. Preliminary observations of frog filariasis in New Jersey. *Bull. Wildlife Disease Assoc., 5:* 342-347.

CRANS, W. J., 1970. The blood feeding habits of *Culex territans* Walker. *Mosquito News, 30:* 445-447.

DAYKIN, P. N., 1967. Orientation of *Aedes aegypti* in vertical air currents. *Can. Ent., 99:* 303-308.

DAYKIN, P. N., KELLOGG, F. E. & WRIGHT, R. H., 1965. Host-finding and repulsion of *Aedes aegypti. Can. Ent., 97:* 239-263.

DOW, R. P., REEVES, W. C. & BELLAMY, R. E., 1957. Field tests of avian host preference of *Culex tarsalis* Coq. *Am. J. trop. Med. Hyg., 6:* 294-303.

DOWNE, A. E. R., 1960. Blood-meal sources and notes on host preferences of some *Aedes* mosquitoes (Diptera:Culicidae). *Can. J. Zool., 38:* 689-699.

FLEMINGS, M. B., 1958. A field study of the host biting preference of *Culex tritaeniorhynchus* Giles and other mosquito species of Japan *Proc. 45th Mtg. N. J. Mosq. Exterm. Assoc.:* 89-97.

FORATTINI, O. P., LOPES, O. de S. & RABELLO, E. X., 1968. Investigacoes sobre o comportamento de formas adultas de mosquitoes silvestres no Estada de Sao Paulo, Brazil. *Rev. Saude publ. S. Paulo, 2:* 111-173. (Abstract in *Rev. appl. Ent. (B), 58:* 388.)

GARCIA, R. & JEFFERY, J., 1969. Observations on mudskippers (*Periophthalmus* spp.) as hosts for mosquitoes (*Aedes (Rhinoskusea)*) in a mangrove swamp. *Proc. Seminar on Filariasis and Immunology of parasitic Infections and Laboratory Meeting, Singapore, Malaysian Soc. Parasit. trop. Med.:* 244-245.

GIGLIOLI, M. E. C., 1965. The influence of irregularities in the bush perimeter of the cleared agricultural belt around a Gambian village on the flight range and direction of approach of a population of *Anopheles gambia melas. Proc. XIIIth Int. Congr. Ent.:* 757-758.

GILLIES, M. T., 1967. Experiments on host selection in the *Anopheles gambiae* complex. *Ann. trop. Med. Parasit., 61:* 68-75.

GILLIES, M. T. & DE MEILLON, B., 1968. The Anophelinae of Africa South of the Sahara (Ethiopian Zoogeographical Region). *Publ. S. A. Inst. med. Res.,* no. 54.

GILLIES, M. T. & SNOW, W. F., 1967. A CO_2-baited sticky trap for mosquitoes. *Trans. R. Soc. trop. Med. Hyg., 61:* 20.

GILLIES, M. T. & WILKES, T. J., 1965. A study of the age composition of populations of *Anopheles gambiae* Giles and *Anopheles funestus* Giles in north-eastern Tanzania. *Bull. ent. Res., 56:* 237-262.

GILLIES, M. T. & WILKES, T. J., 1969. A comparison of the range of attraction of animal baits and of carbon dioxide for some West African mosquitoes. *Bull. ent. Res., 59:* 441-456.

GILLIES, M. T. & WILKES, T. J., 1970. The range of attraction of single baits for some West African mosquitoes. *Bull. ent. Res., 60:* 225-235.

HADDOW, A. J., 1945. The mosquitoes of Bwamba County, Uganda, II. Biting activity with special reference to the influence of microclimate. *Bull. ent. Res., 36:* 33-73.

HARRIS, P., RIORDAN, D. F. & COOKE, D., 1969. Mosquitoes feeding on insect larvae. *Science, N.Y., 164:* 184-185.

HAYES, R. O., 1961. Host preference of *Culiseta melanura* and allied mosquitoes. *Mosquito News, 22:* 182-185.

HOCKING, B., 1971. Blood-sucking behaviour of terrestrial arthropods. *Ann. Rev. Ent., 16:* 1-26.

HOCKING, B. & KHAN, A. A., 1966. The mode of action of repellent chemicals against blood-sucking flies. *Can. Ent., 98:* 921-931.

IYENGAR, R., 1962. The bionomics of salt-water *Anopheles gambiae* in East Africa. *Bull. Wld Hlth Org., 27:* 223-229.

KUPRIYANOVA, E. S. & VOROTNIKOVA, L. M., 1967. K ekologo-biologicheskoi kharakteristike populatsii *Culex pipiens pipiens* L. v Podmoskv'e. *Med. Parasit., Moscow, 36:* 216-224.

LINDQUIST, A. W., IKESHOJI, T., GRAB, B., De MEILLON, B. & KHAN, Z. H., 1967. Dispersion studies of *Culex pipiens fatigans* tagged with ^{32}P in the Kemmendine area of Rangoon, Burma. *Bull. Wld Hlth Org., 36:* 21-38.

MATTINGLY, P. F., 1969. *The biology of mosquito-borne disease.* London: George Allen & Unwin.

MENON, M. A. U. & TAMPI, M. R. V., 1959. Notes on the feeding and egg-laying habits of *Ficalbia* (*Mimomyia*) *chamberlaini,* Ludlow 1904, (Diptera, Culicidae). *Indian J. Malar., 13:* 13-18.

MOROZOV, V. A., 1965. Komar *Culex pipiens* L., pitayushchiyesya krova cheloveka (v okrestnostyakh Krasnodara). *Med. Parasit., Moscow, 34:* 24-29.

MURPHEY, F. J., BURBUTIS, P. B. & BRAY, D. F., 1967. Bionomics of *Culex salinarius* Coquillett. II. Host acceptance and feeding by adult females of *C. salinarius* and other mosquito species. *Mosquito News, 27:* 366-374.

REID, J. A., 1961. The attraction of mosquitoes by human or animal baits in relation to the transmission of disease. *Bull. ent. Res., 52:* 43-62.

SNOW, W. F., & WILKES, T. J., 1970. The height and direction of non-host oriented flight, and the age, of *Anopheles melas* in relation to wind direction in the Gambia, West Africa. *Trans. R. Soc. trop. Med. Hyg., 64:* 477.

TEMPELIS, C. H., HAYES, R. O., HESS, A. D. & REEVES, W. C., 1970. Blood-feeding habits of four species of mosquito found in Hawaii. *Am. J. trop. Med. Hyg., 19:* 335-341.

VINCKE, I. H., 1954. Natural history of *Plasmodium berghei. Indian J. Malar., 8:* 245-256.

WHARTON, R. H., 1962. The biology of *Mansonia* mosquitoes in relation to the transmission of filariasis in Malaya. *Bull. Inst. Med. Res. Malaya,* No. 11. Kuala Lumpur.

DISCUSSION

M. W. Service

What is the shape of the three dimensional host stream, especially with regard to vertical distribution of the attractants such as CO_2 and host odour?

M. T. Gillies

On the basis of physicists' findings on the shape of gas-clouds one would expect the cross-section of the host-stream to be very approximately symmetrical; that is to say, extending upwards from the ground as far as the lateral spread.

J. R. Busvine

The contrast between cave-dwelling mosquitoes (relying on convective and humidity cues to locate hosts) and savannah types (relying primarily on odour at a distance) reminds one of the contrast between blowflies and houseflies. The former depend on odour at a distance (and have antennae well supplied with sense organs) while houseflies, living indoors near to food sources, rely mainly on visual clues and aggregating habits.

J. S. Kennedy

I did not follow the question of small host feeders being distracted by CO_2 plumes since these would always have host odour as well.

M. T. Gillies

The point is that the CO_2 plume from a large animal, or especially from a group of them, might well be longer than the odour plume from a small one.

D. A. Turner

Is there any correlation between the evolutionary development of mosquito genera and the degrees of host-specificity shown? In tsetse flies, which show much less specificity than mosquitoes generally, there seems to be a certain relationship between phylogenetic development and specificity in feeding habits. For instance, members of the *fusca* group are more catholic feeders than the members of the more recent *morsitans* group while the members of the *palpalis* group are intermediate in both respects. There are, however, many exceptions and anomalies.

M. T. Gillies

It's true that among the most primitive mosquitoes, such as in *Toxorhynchites* or *Malaya*, instances of non-blood sucking habits occur, or of amphibian feeding as in *Ficalbia*. But in general no correlation exists between choice of host and evolutionary development.

A. W. R. McCrae

Dr Gillies has pointed out that in open environments mosquitoes tend to fly upwind at levels near to the ground, but downwind at higher levels (perhaps conforming to Johnston's "boundary levels"). One question arising from this is whether samples of mosquitoes taken at different levels show different proportions of the various gonotrophic stages. With respect to movement away from development sites to feeding sites and a return trend of movement after blood-feeding, one might expect a higher proportion of post-feeding mosquitoes at the higher levels. Has this been found to be so?

M. T. Gillies

We have very little evidence on this since our flight-traps have caught very few fed mosquitoes. There is some suggestion of down-wind flights by mosquitoes from the feeding sites at low levels.

S. M. Omer

I may add that up-wind flight by mosquitoes in the absence of host stimuli is dependent on the "general physiological state" of the insects, gonotrophic condition is only one component in the physiology of the behaviour.

D. S. Bertram

In Gambia I have seen mosquito species to have specific habits, to land on particular parts of a bullock—one species at the hooves, another on the neck, a third more generally along the belly. Is this a visual or olfactory selection of particular parts of the host body, or something more complex?

M. T. Gillies

One would suggest that this selection is partly visual and partly due to olfactory selection. It's known, for instance, that convection currents from the body tend to rise from sharp points and angles. It could be that this guides mosquitoes to particular parts of the body.

Host-finding behaviour of tsetse flies

A. G. GATEHOUSE

Department of Zoology and Applied Entomology,
Imperial College of Science and Technology, London

The behaviour of *Glossina morsitans* and *Glossina swynnertoni* through the trophic cycle is described. It is suggested that the favoured hosts of these tsetse flies are likely to be feeding at the times of day when the flies are active, and therefore that most feeding by the tsetse flies occurs while the hosts are also feeding. It is also argued that the slow, intermittent movements of grazing, browsing or foraging mammals allow large concentrations of "following" flies to build up around the hosts.

Laboratory experiments have demonstrated that male teneral *G. morsitans* are more responsive to a moving visual stimulus than females, and that a higher level of response by females is obtained if the movement is interrupted. These results do not appear to relate to what is known of the behaviour of teneral *G. morsitans* in the field. However, if the responses of non-teneral *G. morsitans* to visual and olfactory stimuli prove to be similar, they are likely to be important both in relation to feeding behaviour, and to the formation of the "following swarm".

CONTENTS

INTRODUCTION

Tsetse flies owe their notoriety to the fact that they are the vectors of four species of salivarian trypanosomes of African wild mammals, which cause serious disease in man and his domestic animals. Three of these species, *Trypanosoma vivax*, *Trypanosoma congolense* and *Trypanosoma brucei* are of real importance, and the latter has two subspecies *T. b. rhodesiense* and *T. b. gambiense*, which cause acute and chronic infections, respectively, in man. There are 34 known species of *Glossina*, most of which play no part in disease transmission. In general the disease of cattle, "nagana", and Rhodesian sleeping sickness in man are transmitted by the *morsitans* group flies *G. morsitans*, *G. swynnertoni* and *G. pallidipes*, and Gambian sleeping sickness by the *palpalis* group of riverine or lacustrine flies, *G. palpalis* and *G. fuscipes*.

Since the beginning of the colonial era in Africa efforts have been directed to eliminating these diseases, by the eradication of the trypanosomes by chemical therapy and prophylaxis, or of the vector by a variety of control measures. In his recent book Ford (1971) has pointed out that, while these approaches are obviously impracticable on a continental scale they may be inadvisable or even dangerous when applied to limited areas. Any tolerance of infection based on ecological and physiological factors may be lost, and unless expensive eradication programmes can be maintained indefinitely, men and cattle may be exposed to disastrous resurgence of the diseases. He argues that an immediate objective should be to achieve a better understanding of the causes of epidemics and epizootics and "the development of a sound epidemiology of tsetse born diseases".

It has been known since early this century (e.g. Fiske, 1920), and serological identification of blood meals (Weitz, 1963) has confirmed, that many tsetse species show a distinct hierarchy of host preference and that domestic animals are not always, and man is never at, or even near, the top of this hierarchy. It is, therefore, obvious that knowledge of the factors determining the choice of hosts by tsetse, and of the circumstances under which these preferences may break down and so increase the probability of infection of man or cattle, is central to the epidemiology of the infections. Fiske (1920) clearly stated the necessity of investigating the relationship between tsetse and its hosts, and yet this is still an area of comparative ignorance in the very large cumulation of knowledge of the biology of the genus *Glossina*.

Tsetse flies are rare in the sense that "few occur in the area which a man can hope to search in the course of a day" (Glasgow, 1963). Population estimates quoted by this author (which he considers to represent high rather than typical levels) from studies of *G. morsitans morsitans* (Harley, 1958) and *G. fuscipes fuscipes* (Glasgow, 1953, 1954) range from 0.05-1.0 male flies per 100 sq. yards for the former and 0.2-2.0 for the latter. These figures refer to non-teneral males only (that is, males that have had at least one blood meal). There are probably about twice this number of females. These very low densities and the fact that man himself constitutes a potential host, raise obvious difficulties for ecologists. Virtually all knowledge of the ecology and behaviour of *Glossina* is derived from samples obtained by small parties of men (sometimes carrying screens or accompanied by bait animals) moving through the bush on "fly-rounds" (Potts, 1930; Ford *et al.*, 1959) and stopping at intervals to catch the tsetse that have been attracted to them. Samples of many species collected in this way are found to be predominantly male (usually more than 80%). The conventional explanation (Lloyd, 1912; Lamborn, 1916; Fiske, 1920) is that males show a "sexually appetitive" response to moving objects long before they will make any attempt to feed. They fly towards the object but do not usually land on it and this results in a concentration of males around potential hosts—the "following swarm"— which increases the probability of fertilization of young females, which approach to feed. Fly-round catches are therefore made up of these males and a small number (man is not a favoured host) of hungry flies of both sexes coming to feed. If only those flies which attempt to feed on the men are counted the sex ratio approaches unity.

The bias towards males has resulted in some serious misconceptions. For example, on the basis of fly-round sampling, it had been accepted that populations of *G. morsitans* were largely confined to dense riverine vegetation in the dry season in Rhodesia. However Pilson & Pilson (1967) used a sampling technique (the catch from a tethered bait-ox) which provided a more representative sample of hungry tsetse flies. They showed that the general population remained distributed through the savanna woodland throughout the year, and that it was only males which tended to aggregate in the riverine vegetation in the dry season. Conclusions drawn from biased samples of this sort have often had a major influence on the planning of control measures, particularly the selective clearing of vegetation. Thus, as Bursell (1970) says, the work of Pilson & Pilson raises "the very serious question whether investigations of tsetse ecology have not been bedevilled throughout their history by a vicious sampling error almost reaching the dimensions of a practical joke". It is clear, as fly-round sampling methods are unlikely to be superseded in the immediate future, that interpretation of results depends largely on an understanding of the tsetse-host relationship.

The discussion that follows has been deliberately confined to *G. morsitans* and *G. swynnertoni.* These two species occupy similar habitats in savanna woodland and their behaviour is known to be similar.

BEHAVIOUR THROUGH THE TROPHIC CYCLE

Tsetse are capable of flying at speeds of the order of 15 miles/h. However, in spite of their impressive powers of flight, the best available estimate of their rate of dispersal is only 200 yards per week (Jackson, 1940, 1944, 1948; Glasgow, 1963). The implied conclusion, that tsetse populations tend to be local, is supported by Bursell's (1966) finding that samples taken from areas separated by only a few miles differ consistently in their nutritional state (size and fat content). This evidence is in accordance with the view that tsetse spend the greater part of their life at rest. Observations in the laboratory (Bursell, 1957; Brady, 1970) and in a large field cage (Dean, Williamson & Phelps, 1969) show that the flies move infrequently. Tentative conclusions based on studies of the fat content of teneral flies and the amino-acid content of flies caught on fly-rounds and bait animals, suggest that hungry male flies may spend in the order of 10-15 min in flight per day (Bursell, 1970).

Teneral flies

Activity in teneral (unfed) *G. morsitans* and *G. swynnertoni* appears to be minimal in the first 24 h after emergence, but increases progressively in the second and third days, by which time, under favourable conditions, most flies have probably taken their first blood meal (Jackson, 1946; Bursell, 1959). As they have to rely for their survival entirely on fat and water reserves carried through the pupal stage, the location of a host is an urgent priority for teneral flies of both sexes.

Non-teneral males

The changes in the behaviour of *G. swynnertoni* through a trophic cycle were investigated by Bursell (1961a), who divided the cycle into four phases on the basis of evidence obtained from the amount of residual blood meal and the fat content of flies caught by different sampling methods. Ford (1971) gives an account of the behaviour of male *G. morsitans* based on Bursell's conclusions, and an outline of this description, which represents the accepted view, will be given here.

After engorging, the insect flies laboriously to a nearby tree, failing to gain much height, and remains inactive during the initial phase of digestion in which the blood meal is concentrated by the primary excretion of water. It may then move to a more suitable resting site (e.g. into the canopy of the tree—Pilson & Pilson, 1967). Here it probably rests except for movements to take advantage of favourable micro-climates or to avoid predators, while digestion proceeds and fat reserves are laid down (phase 1). Before digestion is complete, probably about 12 h after feeding, spontaneous activity increases and the flies start to respond to moving objects (Bursell, 1957, 1961a; Brady, 1971). This enhances the chances of encountering a potential host, in association with which these flies form a "following swarm" (phase 2). Male *G. morsitans* in a following swarm behind a party of men, fly slowly on a zig-zag course near the ground, keeping a few yards behind. When the men stop, the flies settle on the ground or low vegetation. Young females are intercepted as they approach the host, or after they have landed on it, and mating follows. The duration of an individual's participation in a particular following swarm appears to be limited because the size of swarms behind fly-round parties does not increase indefinitely (Bursell, 1961a). The fly will take part in following swarms without feeding until some threshold is reached beyond which it will respond to stimuli from a suitable host (possibly the focus of the last following swarm) by landing on it and feeding (phase 3). The factors determining this threshold are not fully understood (Bursell, 1961a; 1966; Ford, 1971). If the fly fails to locate a favoured host and its reserves approach complete exhaustion, it enters phase 4, when it approaches any moving object and attempts to feed, irrespective of host preferences and even eventually, whether or not the object is animate. The length of the cycle depends on environmental conditions but in the hot dry season, is slightly less than three days for *G. swynnertoni* and rather longer in *G. morsitans* (Jackson, 1954).

Males do not become potent until 4-7 days after emergence (Saunders, 1970), and the flight musculature is not fully developed until the second or third blood meals (Bursell, 1961b). There is some evidence that older males may cease to participate in following swarms (Ford, 1969).

Non-teneral females

Very much less is known of the behaviour of females during the trophic-cycle. It is believed that they are even less active than males and that, except for escape reactions, movements to favourable micro-climates and the search for suitable sites for larvi-position, they probably remain at rest between feeds. It is supposed that they only approach hosts in order to feed, and only appear in fly-round catches when their nutritional reserves are exhausted (as tenerals or as phase 4 non-tenerals). However they do appear in equal numbers with males in catches of flies attempting to feed off a tethered bait animal (Pilson & Pilson, 1967) or a "resting" man (Jack, 1941).

TSETSE BEHAVIOUR IN RELATION TO HOST BEHAVIOUR

In the laboratory *G. morsitans* shows a distinct diurnal pattern, with peaks in morning and late afternoon, both of spontaneous activity (Bursell, 1957; Brady, 1970) and of responsiveness to a moving object (Brady, 1971), although in the field such patterns of activity are modified by environmental factors. Thus feeding activity in *G. morsitans* is confined to times when the air temperature is above 18°C. Field studies in Rhodesia show a tendency for peaks in the rate of attack on a tethered ox to occur in the morning and (more uniformly) in the late afternoon through most of the year, and in a variety of vegetation types (Pilson & Pilson, 1967). An exception is in the wet season (December to April) when the attack is maintained throughout the day.

A major difficulty encountered in any attempt to relate the behaviour of tsetse to that of its hosts, is the lack of information on the behaviour patterns of the hosts at the times of day when the flies are active.

The animals most commonly fed upon by *G. morsitans* and *G. swynnertoni* in East Africa were identified by serological examination of blood meals (Weitz & Glasgow, 1956; Weitz, 1963). Almost invariably (but see Glasgow *et al.,* 1958) it is found that between 40% and 50%, and often 70-80%, of feeds of these two species are taken from suids. The vast majority of these are from warthog (*Phacochoerus aethiopicus*). Weitz & Glasgow (1956) classify warthog and rhinoceros as "animals always bitten" when available to the tsetse. The next most common sources of blood are the ruminants, including roan antelope, buffalo, kudu, bush buck, reed buck, and giraffe, which are classed as "commonly bitten". This category also includes elephant and bush pig. One common characteristic of the behaviour of these favoured hosts, with the possible exception of roan antelope, is that they tend to feed in the savanna woodland rather than in the open grassland of vleis and glades. The populations of many of them are also essentially local, and some (including warthog) are repetitive in their habits, tending to use the same resting places, waterholes and game tracks for long periods.

In very general terms, the daily activity of warthog and these larger ungulates may include a visit to a source of water and probably, at least in the hot season, a period of quiescence in dense shade (or in their holes in the case of warthog) in the middle of the day (Harrison, 1936). The remainder of the day is spent grazing, browsing or foraging. A conspicuous feature of the feeding behaviour of these animals is that they move through the woodland slowly, halting for periods from a few seconds to more than a minute, every few paces. It seems reasonable to suppose that it is this pattern of host behaviour that is most often encountered by the tsetse, because, in the middle of the day when the hosts may lie up in the shade, the flies are also likely to be inactive. In the cool and wet seasons when feeding by tsetse occurs from late morning onwards (Pilson & Pilson, 1967) the hosts probably also feed sporadically throughout the day. If tsetse flies do normally encounter their hosts while they are feeding and moving intermittently in this way, it has some interesting implications both for the behaviour of flies in the following swarm, and for feeding behaviour.

It has already been mentioned that when a potential host moves continuously with only occasional short pauses (as occurs on a fly-round), individual flies in the following swarm appear to maintain contact for only a limited period of time, and this conclusion is supported by the evidence that suggests that tsetse do not spend more than about 15 min in flight a day (see above). Under these circumstances the following swarm would be expected rapidly to reach a state of equilibrium in which recruitment is balanced by the numbers left behind. Field observations of following swarms on fly-rounds suggest that this does happen (Bursell, 1961a). However, if the focus of the following swarm is moving only intermittently (for example, a browsing animal), the situation is different. As before, the flies would be attracted by the host's movement and would settle out on nearby vegetation when it stopped. But as the movements are only intermittent and the host only moves a few paces at a time, the fly is likely to be able to maintain contact with a feeding host, or group of hosts, for considerable periods, without remaining in flight for more than a fraction of that time. More flies would be recruited as the animals moved through the bush so that large concentrations would be built up. Harrison (1940) observed large aggregations of fly around various hosts (particularly warthog) but, presumably because of the shyness of the host animals, there are no other reports for *G. morsitans* or *G. swynnertoni* comparable to those of Fiske (1920) who described seeing large concentrations of *G. fuscipes,* showing no signs of feeding, around relatively tame sitatunga and *Varanus* on depopulated islands in Lake Victoria. The formation of large and persistent concentrations of males around hosts would be consistent with the observed fact that non-teneral females are very rarely found to be unfertilized (Vanderplank, 1947). A highly efficient means of ensuring early fertilization is particularly important to a species whose population density and rate of reproduction are so low.

It also follows that much of the feeding by tsetse is likely to take place while the hosts are also feeding. Very little is known of the conditions which are most favourable to tsetse feeding activity, but two factors have been thought to be particularly important. Bursell (1961a) has suggested that it is important that the host should be relatively quiescent. However the evidence he mentions—that the female proportion is greater in "standing catches" than "moving catches" (e.g. Jack, 1941; Jackson, 1930; Bursell, 1961a) and that the female proportion may be regarded as a criterion of the effectiveness of a stimulus situation to elicit feeding behaviour—does not seem to justify such a conclusion. An adequate test of this hypothesis would be difficult to apply and there do not appear to have been any attempts to compare the numbers of flies caught actually feeding on a stationary host with the numbers caught feeding on a moving host under comparable conditions. It has also been suggested that shade may be important (Isherwood and Southon quoted by Bursell, 1961a). Glasgow (1961) showed that, if two men were seated side by side, one with his legs in sunlight and the other with his legs in shade, many more *G. swynnertoni* of both sexes attacked the shaded legs. Whether or not they influence feeding behaviour, both these conditions (low light intensity and relative quiescence) are likely to be met while hosts are browsing or grazing in savanna woodland. The animals would be in shade much of the time and their movements would be slow and intermittent.

If these arguments are valid it would seem that much, if not most of tsetse feeding activity occurs while the hosts are also feeding, and that the slow and intermittent movements of those feeding hosts also allow large and relatively persistent concentrations of flies to build up around them. Of course, as the results of catches on tethered bait animals suggest, some feeding probably occurs while the hosts are lying up. It is however worth emphasizing that a host should never be thought of as motionless, because even a recumbent animal makes constant movements of its head or tail.

RESPONSES OF TENERAL *G. MORSITANS* TO VISUAL AND OLFACTORY STIMULI IN THE LABORATORY

The results of recent laboratory experiments have demonstrated substantial differences between teneral male and female *G. morsitans* in their responses to visual and olfactory stimuli and these differences may have some bearing on the conclusions outlined above (Gatehouse, 1972a).

The experiments were carried out in a wind tunnel which produced a near laminar flow of air of 0.4 m/s in a large flight chamber measuring 1.8 m long by 1.2 m wide by 1.5 m high (Fig. 1). The tunnel airflow was maintained at 25° C and 50-60% RH. An airstream containing olfactory stimuli from a host could be released from a point at the centre of the upwind end of the chamber and this contaminated a discrete cone of the tunnel airflow through the flight chamber. The flies were contained in a metal tube (12.5 cm diameter and 45 cm long) positioned directly downwind from the stimulus airstream release point, and opening into the chamber through a netting funnel mounted in the downwind netting end screen. All flies in the tube or on the inside of the netting funnel were subjected to the olfactory stimulus. The distance between the front of the tube and the upwind end of the chamber was 2.5 m. The visual target was a 30 cm square of matt black perspex which was mounted on horizontal runners and could be driven across the upwind end of the chamber by a servo motor. It passed directly in front of the stimulus airstream release point, and out of the flight chamber at each end of its traverse. The olfactory stimuli were from an Ayrshire bullock which was kept in a stall under a polythene sheet canopy. The stimulus airstream was pumped from the apex of this canopy and released into the wind tunnel at 5 l/min and at the ambient room temperature of 25° C (its relative humidity was not controlled). The airstream therefore contained components from the calf's body surface, exhalant breath and excreta.

The tests were conducted using teneral (unfed) flies 46-60 h after emergence and in the "late afternoon" of their 12 h light : 12 h dark cycle. For 4-5 h precautions were taken to ensure minimal visual and mechanical disturbance. Flies were then subjected to the stimuli while they were at rest in the tube or in the netting funnel (90% of those tested were in the tube). They were, however, free to fly out into the chamber.

In all experiments the flies were subjected to a visual stimulus consisting of one or two traverses of the target at the beginning of each of two successive minutes. The odour airstream was released for 2½ min, starting 30 s before the target first appeared, and all insects flying into the flight chamber in the 2½ min observation were counted.

In the first experiment (Fig. 2A) the target was driven across the flight chamber, and then back again immediately, at 0.25 m/s at the beginning of each of the two minutes. When a control airstream from the calf shed, but with

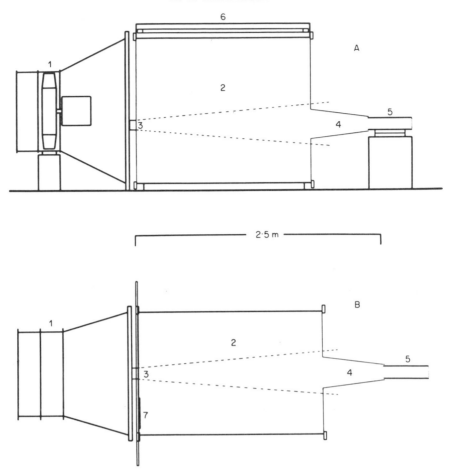

Figure 1. Windtunnel apparatus. The airflow is from left to right at 0.4 m/s. The interrupted lines give an approximate indication of the extent of the tunnel airflow contaminated by the olfactory stimuli. A. Sagittal section. B. Horizontal section in the plane of the olfactory stimulus release point. The visual target (a 30 cm square of matt black perspex) mounted on horizontal runners, is driven across the flight chamber by a servo-motor. It passes immediately in front of the olfactory stimulus release point, and out of the wind tunnel at each end of its traverse.

1, Fan assembly; 2, flight chamber; 3, olfactory stimulus release point; 4, netting funnel opening through downwind netting end-screen; 5, fly release tube; 6, lights; 7, visual target.

no calf present, was released 21% of the males responded. This level of response did not differ significantly from that of males and females when the stimulus airstream was switched off and so in subsequent controls no airstream was released. The release of calf odour did not influence the response of females to this visual stimulus, but it did increase to 47.4% ($P < 0.01$) the proportion of males responding.

The target speed was reduced to 0.09 m/s in the second experiment which was otherwise identical to the first (Fig. 2B). Once again the female response was low whether or not the calf odour was released. However the male response reached 50% even with no calf odour, and this did not differ significantly from the response with calf odour ($P > 0.10$). The flies in the netting funnel or in

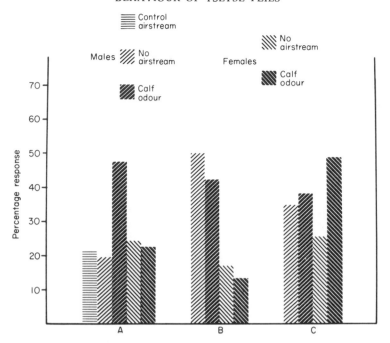

Figure 2. Responses of teneral *G. morsitans* to visual stimuli, with and without the simultaneous release of olfactory stimuli from a bullock. A. Two traverses of visual target at 0.25 m/s each minute. B. Two traverses of visual target at 0.09 m/s each minute. C. One traverse of the visual target at 0.09 m/s with a 5 s halt each minute.

the very front of the release tube had an unrestricted view of the whole traverse of the target, but those at the downwind end of the tube may have been able to see as little as 30% of its traverse, and only 70% of its area when both leading and trailing edges were in view. There was, therefore, a gradation in the intensity of the stimulus, in terms of its duration and area, according to the position of the fly in the tube, and the results suggest that responses of the order of 50% represent a maximum in this system.

In the third experiment (Fig. 2C) the visual stimulus at the beginning of each minute consisted of only a single traverse (in a different direction each time) at 0.09 m/s, but the target was stopped for 5 s just after it had passed the halfway point. The male response was high with or without the release of calf odour, and although slightly lower, did not differ significantly from that in the previous experiment. It was, however, significantly higher than the response in the two controls in the first experiment ($P < 0.05$). The response of females with no olfactory stimulus was comparable to that obtained in previous experiments ($P > 0.10$) but was increased to 48.8% ($P < 0.05$) when the calf odour was released. It seems that this high female response can only be attributed to the interruption in the movement of the target. All other features of the visual stimulus likely to contribute to its intensity have been diminished, since the target entered and left the insects' field of vision half as many times, and remained in view for a shorter period of time than in the second experiment.

In the controls of all these experiments, when no odour was released, very few of the flies (amounting to only 4.8% of the total number responding) took off when the visual target was outside their field of view, and none did so before its first appearance. In contrast 28.6% of take-offs, of which one-third were before the target appeared, occurred with the target outside the chamber in the presence of the olfactory stimulus. Almost all the flies which took off while the target was in view flew straight to it turning away at the last minute or settling on it, or on the netting near by. Of those which flew while the calf odour was being released but with the target outside the chamber, some flew straight upwind but the flight paths of others appeared to bear no relation to the odour release point. It was also noticed that, in the presence of the olfactory stimuli, those flies that did respond remained in flight longer than did those in experiments which did not involve the release of calf odour.

The results of this last experiment also demonstrated that the response of males to visual stimuli tended to be more immediate than that of females. Ninety per cent of the male flights, as opposed to only 64.5% of those of females occurred before the second visual stimulus was presented. Compared with only 15.4% of male flights, 38.5% of the female flights which occurred while the target was in the flight chamber, started during the 5.0 s halt in its traverse. Of the remainder the majority of male responses were observed before the target stopped, whereas females tended to fly as the target was moving out of the chamber after the halt.

When the flies were not presented with any visual stimulus, their response to calf odour was low. In a further experiment the procedure followed was exactly as before, except that observations were made for 3 min before the olfactory stimulus airstream was switched on, and for the following 3 min. No flies of either sex took off in the first 3 min, and only 14.3% of the males and 13.7% of females flew into the flight chamber in response to the calf odour. This is approximately the same proportion as responded to calf odour before the first appearance of the visual stimulus in the first three experiments.

In the final experiment, a rectangle of black cartridge paper measuring 45 cm by 20 cm (i.e. with an area equal to that of the moving target) was mounted on each side of the stimulus airstream release point. The procedure was the same as in the last experiment and this meant that the stationary visual target was within the flies' field of view for 20 min before the observations started. Again no take-offs occurred before the odour was released, and only 10% of males and 11.9% of females responded in the subsequent 3 min.

The results of these experiments show that teneral male G. morsitans are more responsive to a moving target than females. The slower speed was the more effective stimulus for males, inducing a high level of response even without the olfactory stimulus. The angular velocity of the target was between 4.8 and 7.5 °/s in the first experiment and 1.7 and 2.7 °/s in the second, according to the fly's position in the release tube or the netting funnel. A high level of response by females was only achieved when the movement of the target was interrupted briefly, and then only when the olfactory stimulus was presented. The final experiment indicates that this response by females depended on both the "moving" and "stationary" components of the visual stimulus. There also appears to be a difference between the sexes in the immediacy of their response to the target, with males taking off sooner after its appearance than females.

The main effect of the olfactory stimuli seems to be to increase the responsiveness of both sexes to visual stimuli. However a small proportion of the flies of both sexes consistently flew into the flight chamber when they were subjected to the olfactory stimulus alone, and there is other evidence which suggests that olfactory stimuli may play a more important part in host location, at least by teneral female *G. morsitans*, than these experiments suggest (Gatehouse, 1972b). Nothing is known of the composition of the olfactory stimuli from the calf. However experiments have shown that the release of uncontaminated outside air at 100% RH and ambient temperature did not influence the response to the target, and there is circumstantial evidence that the presence of carbon dioxide is not important.

It must again be emphasized that these results all refer to teneral *G. morsitans*. The sex ratios of catches of teneral flies on fly-rounds or bait animals usually approach unity, and the exceptions that have been observed do not appear to be explicable in terms of the results of these experiments (e.g. Nash, 1933; Jackson, 1944; Pilson & Pilson, 1967). Any attempts to relate these results to the behaviour of non-teneral flies in the field is obviously premature. However, if the observed differences between males and females in their responsiveness to visual stimuli persist in the non-teneral state, it is likely to have an important bearing on the formation and composition of the "following swarm". It is also interesting to speculate that the slow intermittent movement of hosts characteristic of browsing, grazing and foraging, may prove to be an important component of the stimuli inducing feeding behaviour.

ACKNOWLEDGEMENTS

I thank Dr C. T. Lewis for helpful discussion and Dr J. N. Brady for reading and criticizing the manuscript. Pupae of *G. morsitans morsitans* Westwood were obtained from the Tsetse Research Laboratory, Department of Veterinary Medicine, University of Bristol, and the work was financed by the Overseas Development Administration of the Foreign Office.

REFERENCES

BRADY, J., 1970. Characteristics of spontaneous activity in tsetse flies. *Nature, Lond., 228:* 286-287.
BRADY, J., 1971. Laboratory observations on the visual responses of tsetse flies. *Trans. R. Soc. trop. Med. Hyg., 65:* 226-227.
BURSELL, E., 1957. The effect of humidity on the activity of tsetse flies. *J. exp. Biol., 34:* 42-51.
BURSELL, E., 1959. The water balance of tsetse flies. *Trans. R. ent. Soc. Lond., 3:* 205-235.
BURSELL, E., 1961a. The behaviour of tsetse flies (*Glossina swynnertoni* Austen) in relation to problems of sampling. *Proc. R. ent. Soc. Lond. (A), 36:* 9-20.
BURSELL, E., 1961b. Post-teneral development of the thoracic musculature in tsetse flies. *Proc. R. ent. Soc. Lond. (A), 36:* 69-74.
BURSELL, E., 1966. The nutritional state of tsetse flies from different vegetation types in Rhodesia. *Bull. ent. Res., 57:* 171-180.
BURSELL, E., 1970. Dispersal and concentration of *Glossina*. In H. W. Mulligan (Ed.), *The African Trypanosomiases*. London: George Allen & Unwin.
DEAN, G. J. W., WILLIAMSON, B. R. & PHELPS, R. J., 1969. Behavioural studies of *Glossina morsitans* Westw. using tantalum–182. *Bull. ent. Res., 58:* 763-771.
FISKE, W. F., 1920. Investigations into the bionomics of *Glossina palpalis*. *Bull. Ent. Res., 10:* 347-463.
FORD, J., 1969. Feeding and other responses of tsetse flies to man and ox and their epidemiological significance. *Acta trop., 26:* 249-264.
FORD, J., 1971. *The role of the Trypanosomiases in African ecology*. Oxford: Clarendon Press.
FORD, J., GLASGOW, J. P., JOHNS, D. L. & WELCH, J. R., 1959. Transect fly-rounds in field studies of *Glossina. Bull. ent. Res., 50:* 275-285.

GATEHOUSE, A. G., 1972a. Some responses of tsetse flies to visual and olfactory stimuli. *Nature, Lond.,* *236:* 63-64.

GATEHOUSE, A. G., 1972b. Some responses of *Glossina morsitans* to host odour. *2nd Symposium on Tsetse Fly Breeding and its Practical Application, Bristol, 1971. Trans. R. Soc. trop. Med. Hyg., 66:* 313-314.

GLASGOW, J. P., 1953. The extermination of animal populations by artificial predation and the estimation of populations. *J. anim. Ecol., 22:* 32-46.

GLASGOW, J. P., 1954. *Glossina palpalis fuscipes* Newstead in lakeside and in riverine forest. *Bull. ent. Res., 45:* 563-574.

GLASGOW, J. P., 1961. The feeding habits of *Glossina swynnertoni* Austen. *J. anim. Ecol., 30:* 77-85.

GLASGOW, J. P., 1963. *The distribution and abundance of tsetse.* Oxford: Pergamon Press.

GLASGOW, J. P., ISHERWOOD, F., LEE-JONES, F. & WEITZ, B., 1958. Factors influencing the staple food of tsetse flies. *J. anim. Ecol., 27:* 59-69.

HARLEY, J. M. B., 1958. The availability of *Glossina morsitans* Westw. in Ankole, Uganda. *Bull. ent. Res., 49:* 225-228.

HARRISON, H., 1936. The Shinyanga Game Experiment: a few of the early observations. *J. anim. Ecol., 5:* 271-293.

HARRISON, H., 1940. *The game experiment—Tsetse research report 1935-1938.* Dar-es-Salaam: Government Printers.

JACK, R. W., 1941. Notes on the behaviour of *Glossina pallidipes* and *G. brevipalpis* and some comparisons with *G. morsitans. Bull. ent. Res., 31:* 407-430.

JACKSON, C. H. N., 1930. Contributions to the bionomics of *Glossina morsitans. Bull. ent. Res., 21:* 491-527.

JACKSON, C. H. N., 1940. The analysis of a tsetse fly population. *Ann. Eugen., 10:* 332-369.

JACKSON, C. H. N., 1944. The analysis of a tsetse fly population. II. *Ann. Eugen., 12:* 176-205.

JACKSON, C. H. N., 1946. An artificially isolated generation of tsetse flies (Diptera). *Bull. ent. Res., 37:* 291-299.

JACKSON, C. H. N., 1948. The analysis of a tsetse fly population. III. *Ann. Eugen., 14:* 91-108.

JACKSON, C. H. N., 1954. The hunger cycles of *Glossina morsitans* and *G. swynnertoni* Austen. *J. anim. Ecol., 23:* 368-372.

LAMBORN, W. A., 1916. Second report on *Glossina* investigations in Nyasaland. *Bull. ent. Res., 6:* 249-265.

LLOYD, LL., 1912. Notes on *Glossina morsitans,* Westw., in the Luangwa valley, Northern Rhodesia. *Bull. ent. Res., 3:* 233-239.

NASH, T. A. M., 1933. The ecology of *Glossina morsitans,* Westw., and two possible methods for its destruction. Part I. *Bull. ent. Res., 24:* 107-157.

PILSON, R. D. & PILSON, B. M., 1967. Behaviour studies of *Glossina morsitans* Westw. in the field. *Bull. ent. Res., 57:* 227-258.

POTTS, W. H., 1930. A contribution to the study of numbers of tsetse fly (*Glossina morsitans* Westw.) by quantitative methods. *S. Afr. J. Sci., 27:* 491-497.

SAUNDERS, D. S., 1970. Reproduction of *Glossina.* In H. W. Mulligan (Ed.), *The African Trypanosomiases.* London: George Allen & Unwin.

VANDERPLANK, F. L., 1947. Experiments in the hybridisation of tsetse flies (*Glossina* : Diptera) and the possibility of a new method of control. *Trans. R. ent. Soc. Lond., 98:* 1-18.

WEITZ, B., 1963. The feeding habits of *Glossina. Bull. Wld Hlth Org., 28:* 711-729.

WEITZ, B. & GLASGOW, J. P., 1956. The natural hosts of some species of *Glossina* in East Africa. *Trans. R. Soc. trop. Med. Hyg., 50:* 593-612.

DISCUSSION

J. Ford

The percentage of females in the catch on a fly-round can be increased, as shown some years ago by Welch, by shortening the intervals between stops.

In Landrover catches, the flies outside the vehicle show a low percentage of females but inside the ratio is 50 : 50. Has Dr Gatehouse any explanation?

A. G. Gatehouse

The increased proportion of females in catches with reduction in the length of the intervals between catching stops is most interesting. However the minimum distance travelled between stops in the work you refer to was 25 yards, which is a long way in relation to the sort of intermittent movement I have been talking about.

The explanation I would offer for the 50 : 50 male to female ratio inside the Landrover, is that these are very hungry flies and represent the part of the catch which would have fed, had the target been animate. Catches of flies actually probing men or bait animals on fly-rounds contain the sexes in equal proportions.

D. S. Bertram

The same problem seems also to arise in interpreting the high percentage of females taken in standing, baited traps.

A. G. Gatehouse

Trap catches are extremely difficult to interpret in terms of the behaviour of the fly, and I would rather not attempt to do so

D. A. Turner

Serological investigations of blood meals show suids to be the most favoured hosts of both *Glossina austeni* and *G. morsitans*. Olfactometer investigations have shown, however, that whereas *G. austeni* can be attracted to the odour of guinea-pig, rabbit and man, pig odour is not attractive. This is despite the fact that the relative size, CO_2 output, heat and moisture content of pig and man were very comparable.

A. G. Gatehouse

I think I am right in saying that with the exception of *G. tachinoides*, tsetse are not known to feed readily on domestic pigs. Attempts to use domestic pigs as bait animals on fly-rounds have resulted in poor catches.

P. Boreham

Recent work on the feeding preferences of *Glossina swynnertoni* suggests that buffalo and other bovids are more important as hosts than the earlier literature suggests.

A. G. Gatehouse

This is an important point and the term "favoured host" should only be used in the context of the location being considered.

J. R. Busvine

One would like to think that work was in progress on the immensely difficult subject of the chemical nature of attractive odours for tsetse (and, of course, mosquitoes).

C. A. Wright

While working on the edge of the Victoria Nile in Uganda two of us wearing brown shirts were persistently attacked by tsetse but two others in blue shirts were scarcely affected. Is there any evidence of colour perception in tsetse flies?

A. G. Gatehouse

Some work has been done on this point by Simpson and, later, Moggridge. I think both concluded that there was no evidence that tsetse perceive colour, but they do settle preferentially on surfaces of low reflectance, as do other blood-sucking flies, for example mosquitoes and *Stomoxys calcitrans*.

Behavioural aspects of the life cycle of *Loa*

B. O. L. DUKE

Helminthiasis Research Unit, Institut de Récherches Médicales,
KUMBA, Federal Republic of Cameroon

In the Cameroons rain-forest infections with *Loa loa* in man show diurnally periodic microfilariae which are transmitted by *Chrysops silacea* and *Chrysops dimidiata,* species whose habits bring them into close association with man. Three species of monkey are infected with simian *Loa* sp., which shows nocturnally periodic microfilariae and is transmitted by the crepuscular and nocturnal, canopy-dwelling *Chrysops langi,* and *Chrysops centurionis.*
The human and simian parasites appear to have evolved by adaptive radiation in two separate host-vector complexes. Hybridization between the two parasites is still possible under experimental conditions but in nature, as a result of the behavioural patterns of the microfilariae, their definitive hosts and their vectors, the exchange of parasitic material between the two complexes appears to be minimal and to a great extent self-eliminating.

CONTENTS

INTRODUCTION

In the rain-forest zone of West Cameroon, infections with *Loa loa* exhibiting the usual diurnal microfilarial periodicity are found in about 30% of the human population, and the vectors of the parasite are the day-biting anthropophilic *Chrysops silacea* and *Chrysops dimidiata* (Bombe form). In the forest adjoining human habitations the three most common monkey species, the drill (*Mandrillus leucophaeus*), the putty-nosed guenon (*Circopithecus nictitans martini*) and the mona monkey (*Circopithecus mona mona*) have also been found naturally infected with a form of *Loa* (Gordon, Kershaw, Crew & Oldroyd, 1950). This discovery raised the question as to whether monkeys might constitute a reservoir for the human parasite thereby rendering control measures more difficult to apply. An investigation into the status of the simian *Loa* parasites had therefore to be undertaken.

97

HUMAN- AND SIMIAN-DERIVED *LOA* INFECTIONS IN DRILLS

Monkeys shot in the forest were dissected to determine the infection rate and the average worm load in each species. The adult worms and the microfilariae in blood films were collected and examined morphologically, but apart from their greater somatic size, both adults and microfilariae of the simian parasite appeared to be indistinguishable from the worms described in man. Table 1 gives the infection rates, mean worm loads and the range of dimensions of the parasites found in wild monkeys shot near Kumba. The figures suggest that the parasite in monkeys is best adapted to the drill with a varying spill-over into the two circopithecids.

Table 1. The infection rate with *Loa,* the average worm loads and the dimensions of the parasites in three species of forest-dwelling monkeys

	M. leucophaeus	C. n. martini	C. m. mona
Infection rate (%)	96	24	12
Mean no. of worms/infected monkey	16.6	6.3	2.4
Range of lengths of:			
male worms (cm)	3.6-4.0	3.6-4.0	2.6-3.0
female worms (cm)	8.6-9.0	8.1-8.5	6.6-7.0
microfilariae (μm)	267.5-270	262.5-265	——

Since, apart from size, no clear-cut morphological differences could be detected between the human and simian parasites, the next step was to determine whether the human parasite could be transmitted to monkeys and, if this proved possible, to compare its behaviour in these hosts with that of the natural simian parasite. As *M. leucophaeus* was the most easily obtainable monkey and proved hardy and easy to keep in captivity, all the experiments were done on young specimens of this species. Most of them had been caught when the mother was shot, but some were taken in traps. Before use they were kept in quarantine for six months in screened cages so as to be sure that they were free from natural infections with *Loa*. Following this period some were infected by transplant of live adult simian *Loa* worms taken from freshly-shot wild drills, and others were infected by inoculation of infective larvae dissected out from *Chrysops* which had been fed 9-10 days previously on human carriers of *L. loa* microfilariae.

About a month after the transplant of adult simian *Loa* worms, microfilariae appeared in the peripheral blood of the recipient monkeys, and when counts were made on blood-films, taken hourly throughout the 24 h, it was a considerable surprise to find that the parasites exhibited a clear-cut nocturnal periodicity. By contrast, when the infective larvae of human origin came to maturity in the drill, after a pre-patent interval of some 4-5 months, the microfilariae in the peripheral blood were found to have a typically diurnal periodicity exactly as in man (Duke & Wijers, 1958). These contrasting periodicities, which are shown in Fig. 1A, held true in all the monkeys tested and, moreover, they were maintained after cyclical passage through *Chrysops*.

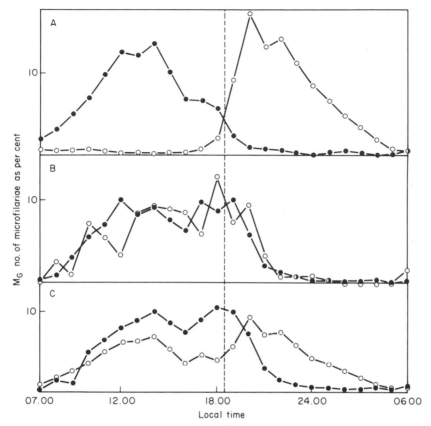

Figure 1. The periodicities of microfilariae of different strains of *Loa* in *Mandrillus leucophaeus.* A. Human strain of *Loa loa* (DD) (●) and the natural simian strain of *Loa* (NN) (○). B. Hybrid microfilariae (DN) emerging from NN females (●) and from DD females (○; drill Sambo). C. Pooled curve for hybrid microfilariae (DN) (●) compared with a theoretical curve for DD and NN microfilariae (○) in equal proportions.

This striking behavioural difference between the natural simian and the natural human parasites when translated to the same experimental host species, coupled with the fact that the worms of the human strain in the drill were considerably smaller (males 2.0-2.5 cm, females 4.0-4.5 cm, microfilariae 222.5-225 μm) than the natural simian parasite, showed that we were dealing with two fundamentally different strains of *Loa* and suggested that there might well be separate vector species of *Chrysops* associated with each host-parasite complex.

LOA INFECTIONS IN *CHRYSOPS* SPP. AND THEIR BEHAVIOUR ON INOCULATION TO DRILLS

For the transmission of simian *Loa* the existence of a night-biting vector *Chrysops* feeding on monkeys sleeping at canopy level was postulated. It was thought that *Chrysops longicornis, C. langi* or *C. centurionis* might be involved in this role for, despite the fact that all three species could be bred as abundantly as could *C. silacea* and *C. dimidiata* from larvae taken in the forest mud, they were virtually never taken on the wing and their biting habits remained a mystery. Moreover *C. centurionis* was known to be a crepuscular and nocturnal feeder in Uganda (Haddow, 1952). To investigate this problem scaling ladders were erected up tall trees in the forest, catching platforms were built at heights of 80-90 feet above ground-level in the canopy layer, and a series of 24-h catches of *Chrysops* was begun using human bait. It soon became apparent (Fig. 2) that *C. silacea* and *C. dimidiata* could be taken during the daylight hours in the canopy as well as at ground-level, but their biting was all but finished by 17.30 hours; in contrast two other species, *C. langi* and *C. centurionis,* were caught biting only at canopy level from about 17.00 hours until 21.00 hours, i.e. at the time and place where monkey troops congregate to sleep (Duke, 1958). The biting curves of the four species are shown in Fig. 2. *C. longicornis* was not taken at all and its biting habits remain unknown in the forest.

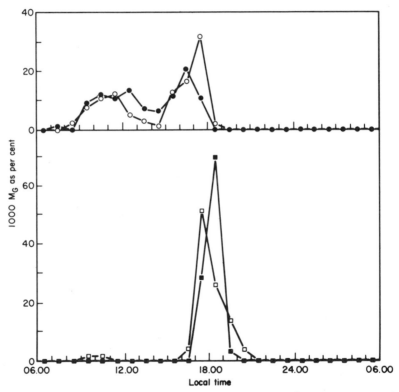

Figure 2. The 24-h biting cycles of four species of *Chrysops* at canopy level in the rain-forest. The hourly biting densities have been computed from the geometric means (multiplied by 1000 to give 1000 MG) of the catches for each hour, and are expressed as percentages of the total 24-h biting density. ○, *C. silacea;* ●, *C. dimidiata* (Bombe form); □, *C. langi;* ■, *C. centurionis.*

The flies caught at the canopy platform in the forest were dissected and it was found that all four species were carrying infections with *Loa* parasites. The next step was therefore to inoculate the "wild" infective larvae from each *Chrysops* species into a series of clean drills. The infections resulting from the larvae carried by *C. langi* (three flies) and *C. centurionis* (one fly) all produced nocturnally-periodic infections with adult worms and microfilariae of large somatic size, while those from wild *C. silacea* (17 flies) and *C. dimidiata* (one fly) produced only diurnally periodic infections with adult worms and microfilariae of small size.

FACTORS INFLUENCING THE INTERCHANGE OF PARASITES BETWEEN HUMAN AND SIMIAN HOSTS

Occurrence of interchange

These findings suggested that wild infections in *C. langi* and *C. centurionis* were all of simian origin and that those from *C. silacea* and *C. dimidiata* were of human origin. However, as it was now known that the human parasite could be transmitted experimentally to the drill, and as *C. silacea* and *C. dimidiata,* the carriers of the human parasite, were known to bite at canopy level, it was decided to investigate the worms found naturally in the wild drills more carefully. In nine trapped drills, which were found to be naturally infected when captured alive, the worms were all of large size and the microfilariae showed a nocturnal periodicity. Similar findings were also recorded in two further drills infected by transplant of worms from wild drills, and in a single example infected by transplant from a naturally infected *C. n. martini* monkey. A search was then made among the worms from wild drills for any of small somatic size whose dimensions might correspond with those of the human parasite when living in experimental drills. Out of 784 worms examined dead only 13 were of dimensions compatible with the human parasite and some of these may well have been immature forms of the simian parasite which had not yet reached full size. However, on one occasion we were fortunate enough to obtain two such worms alive and were able to transplant them into a clean drill. Regrettably, the periodicity curve of the microfilariae in this animal could not be completed for the monkey escaped during the course of the examination and received a fatal injury during recapture, but the figures obtained during the daylight hours before death were strongly suggestive of a diurnal periodicity (Duke & Wijers, 1958). It appears probable therefore that the human parasite is occasionally transmitted to monkeys in nature through the agency of the anthropophilic *Chrysops* species, but it is obviously a rare event and other factors, which will be discussed below, combine to eliminate its chances of further establishing itself in the monkey population. These factors depend partly on the peculiar population dynamics of *Loa* microfilariae in the drill and partly on the behaviour of the hybrid worms formed from crosses between the two strains. The problem has been clarified to some extent by an experimental approach.

Host reactions

In man infections with *L. loa* may exhibit widely different degrees of microfilaraemia from zero up to over 1000 microfilariae per 50 mm^3 of peripheral blood, but once a heavy microfilaraemia is established it is likely to be maintained over long periods and no obvious suppressive mechanism exists affecting the microfilariae. In the drill on the other hand the parasite strikes a different balance with the host's defence reactions (Duke, 1960a). The mean pre-patent interval between the inoculation of infective larvae and the first appearance of microfilariae in the peripheral blood was found to be 147-159 days for the diurnally periodic human parasite in the drill and 135-144 days for the nocturnally periodic simian parasite; these figures suggest that the natural parasite is the better adapted of the two. The microfilariae are born in the loose subcutaneous and inter-muscular connective tissue and their passage into the bloodstream is effected via the lymphatic system. When the microfilariae first enter the bloodstream they accumulate only in the pulmonary circulation and do not begin their periodic invasion of the peripheral blood until some days later. At this stage they can be found abundantly in the lung blood, but films taken from the peripheral capillary bed or from the peripheral veins remain negative. The time interval between birth of microfilariae in the tissues and their first appearance in the peripheral blood can be determined from transplant experiments using adult worms, and is of the order of three weeks. Following their first appearance in the peripheral blood the count of microfilariae at the appropriate time in the 24 h increases fairly regularly for some 9-12 weeks and then falls rapidly over the ensuing 3-6 weeks to level out at a very low density (often less than 10 microfilariae 50 mm^3), which persists throughout the remaining course of the infection. It has been shown that the suppression of microfilariae in the peripheral blood is brought about by the spleen (Duke, 1960b). This organ becomes hypertrophied and covered with most striking granulomata wherein microfilariae are destroyed by macrophages and giant cells amid an eosinophilic cellular reaction. As long as the spleen is intact, and despite the presence of abundant fecund female worms in the tissues and a large reservoir of microfilariae in the lung blood, the density of microfilariae in the peripheral blood remains very low in the monkey host. Once established, the suppressive action of the spleen is sufficient rapidly to destroy any further waves of microfilariae that may arrive in the blood from superinfections by new adult worms regardless of the strain to which they belong. On the other hand when the spleen is removed surgically the suppression of microfilariae is abolished completely and the count in the peripheral blood rises steadily over a period of about a year until it reaches a very high level at which death of microfilariae from old age balances the output of young ones by the adult worms (Duke, 1960a).

Under natural conditions the results of the splenic suppression of microfilariae in monkeys is that, although over 90% of the population may be infected, the numbers of microfilariae available to *Chrysops* feeding on the monkey population are likely to be relatively low. In man on the other hand, there being no splenic or other significant suppressive mechanism known, although a relatively low proportion of persons may show microfilaraemia, a good proportion of those infected will present high concentrations of microfilariae to the feeding flies. This state of affairs is reflected in the infection rates and the number of larvae of *Loa* found naturally in the different species of vector *Chrysops* illustrated in Table 2. *C. silacea* and *C. dimidiata*, which we believe from other evidence to be feeding largely on man, show low infection rates but high mean numbers of parasites per fly. *C. langi* and *C. centurionis*, which are thought to be feeding on monkeys, show high infection rates but much lower numbers of parasites per fly.

Table 2. The infection rate with *Loa* and the geometric mean numbers of parasites per fly carried by wild-caught *Chrysops* of different species

	Percentage infected	Percentage infective	MG no. of parasites per fly
C. silacea	4.7	1.6	79
C. dimidiata	3.0	1.0	81
C. langi	12.6	5.3	26
C. centurionis	31.8	13.6	12

Contact between host and vector

The behavioural evidence so far adduced leads us to think that there is very little interchange of parasitic material in nature between man and monkey but the extent to which interchange can and does take place must be considered further. Humans, whose hairless skins provide a ready source of blood-meals for *Chrysops*, are constantly exposed and frequently bitten by day-biting *C. silacea* and *C. dimidiata*, the more so since these species are attracted down from the canopy by the smoke from wood fires (Duke, 1955), a pattern of behaviour which brings them into close contact with man and his habitations.

By contrast the chances of *C. silacea* and *C. dimidiata* obtaining blood-meals from monkeys during the daylight hours are probably slender, for the hairless portions of the monkey body, from which *Chrysops* can readily feed, are small in extent, and we have observed in the forest that captive monkeys when awake almost invariably catch and eat any attacking *Chrysops* that hover round them before they have a chance to bite. Furthermore, the almost invariable nocturnal periodicity of the natural simian microfilariae, together with the suppressive action of the spleen, will mitigate against the ingestion of many larvae by any day-biting fly which may succeed in taking a full blood-meal from a monkey. As a result it is unlikely that in any *C. silacea* or *C. dimidiata* which bite man after taking a previous blood-meal from a monkey will in fact transmit to him any nocturnal simian larvae.

The difficulty that *Chrysops* have in obtaining blood from monkeys by day is readily overcome at night. The sleeping monkey, squatting on the branch of a tree and no longer alert, presents a very much easier target for those *Chrysops* which continue biting after dusk. The crepuscular *C. langi* and *C. centurionis* both tend to approach the host from below and will often settle on a branch before crawling on to the host's skin. Their habit of biting is superbly adapted to a stealthy approach to the hairless buttocks of a monkey overhanging the branch on which it sleeps, and the simian parasite is obviously efficiently transmitted by these species among the monkey population. On the other hand under normal circumstances the exposure of man to *C. langi* and *C. centurionis* must be virtually zero. Apart from fly-boys stationed there for the purpose, humans are not in the habit of sitting up in the high forest canopy at dusk and during the night, and although the flies would readily bite them if they placed themselves in this position, their natural exposure to the vectors of the nocturnally periodic strain of simian *Loa* is practically nil. No nocturnally periodic infection with *Loa* has ever been reported in man despite frequent examination of blood-films taken at night, nor have attempts to infect volunteers with the simian parasite, either by transplant of adult worms or by inoculation of infective larva, ever succeeded. It is even possible, therefore, that the human race may have some natural resistance to simian *Loa*.

Chances of hybridization

From what has gone before it appears that the extremely limited transference of parasitic material between man and monkey is largely one way. The human parasite is sometimes transmitted to monkeys in nature and the probable fate of such worms must now be considered. Since most monkeys that receive any infective larvae of the human strain will already be infected with their own nocturnal parasite, it has been necessary to investigate the possibilities of interbreeding between the two strains and to study the effects of this on parasite interchange and transmission (Duke, 1964). To do this, two clean experimental monkeys were inoculated, one with infective larva of the diurnal and the other with infective larvae of the nocturnal strain of *Loa*. Both monkeys were killed 70 days later and the young developing parasites, which were identifiable as to sex but were still not mature or fertilized, were removed at autopsy. The virgin nocturnal females (NN) were then transplanted to a clean drill along with diurnal (DD) males, and a second drill was transplanted with virgin diurnal females and nocturnal males. Both these crosses produced microfilariae which were considered to be hybrids (DN), and although their numbers were scanty in the blood they exhibited a characteristic periodicity which, while being predominantly diurnal, showed a considerable extension of the curve into the early hours of the night (Fig. 1B and C).

The hybrid microfilariae were capable of developing in *Chrysops* and, when the infective larvae therefrom were inoculated into clean drills, adult worms (DN) were produced of large somatic size similar to the nocturnal simian strain. The hybrid adults mating among themselves produced microfilariae of complex periodicity, but a study of the length and periodicity of the microfilariae, together with the· results of back-crossing (DN) hybrid males with virgin females of both parent strains in turn, suggested that the two strains were segregating on simple Mendelian lines with regard to periodicity and somatic size.

Further experiments were then done in which both parent strains were allowed to develop simultaneously from infective larvae in the same drill, thus imitating the state of affairs that might occur in nature. There was evidence that some cross-matings had taken place, although other female worms had been fertilized by males of the same strain.

From these experiments we may hypothesize as to what may be happening under conditions of natural transmission. On the rare occasions when infective *C. silacea* or *C. dimidiata* carrying the diurnal human strain of *Loa* succeed in biting a monkey, it appears that any diurnal worms which develop are most likely to be fertilized by the nocturnal male worms already present in the host. Any male diurnal worms which develop are by the same token unlikely to find female nocturnal worms which have not already been fertilized. Furthermore, if it is recalled first that the product of pure diurnal matings or of hybrid matings will produce microfilariae of predominantly diurnal periodicity, second that hybrid matings are not productive of large numbers of microfilariae, and third that the suppressive action of the spleen will tend to keep the numbers of microfilariae very low in the peripheral blood, then it appears improbable that any diurnal worms introduced will succeed in establishing themselves for satisfactory onward transmission by the main night-biting simian vectors *C. langi* and *C. centurionis*.

In conclusion it can be seen that a multitude of behavioural factors affecting parasite, vector and host have contributed to the divergent evolution of human and simian *Loa* which, although they can still produce fertile crosses, must be approaching a separation of true specific rank. The two parasites appear to have undergone an adaptive radiative evolution in separate host-vector complexes, the one involving man with *C. silacea* and *C. dimidiata,* and the other involving the drill and some other monkey species with *C. langi* and *C. centurionis.* Between these two complexes parasite interchange is still possible but occurs very rarely in nature.

SUMMARY

An account is given of the behavioural aspects of the human and simian strains of *Loa* in the experimental host *Mandrillus leucophaeus* and in various vector species of *Chrysops*. The factors that have led to the separation of the two host-parasite-vector complexes are discussed.

ACKNOWLEDGEMENTS

Figures 1 and 2 are reproduced by kind permission of the Editors of the Annals of Tropical Medicine and Parasitology, Liverpool.

REFERENCES

DUKE, B. O. L., 1955. Studies on the biting habits of *Chrysops*. II. The effect of wood fires on the biting density of *Chrysops* silacea in the rain-forest at Kumba, British Cameroons. *Ann. trop. Med. Parasit., 49:* 260-272.

DUKE, B. O. L., 1958. Studies on the biting habits of *Chrysops*. V. The biting cycles and infection rates of *C. silacea, C. dimidiata, C. langi* and *C. centurionis* at canopy level in the rain-forest at Bombe, British Cameroons. *Ann. trop. Med. Parasit., 52:* 24-35.

DUKE, B. O. L., 1960a. Studies on loiasis in monkeys. II. The population dynamics of the microfilariae of *Loa* in experimentally infected drills (*Mandrillus leucophaeus*). *Ann. trop. Med. Parasit., 54:* 15-31.

DUKE, B. O. L., 1960b. Studies on loiasis in monkeys. III. The pathology of the spleen in drills (*Mandrillus leucophaeus*) infected with *Loa. Ann. trop. Med. Parasit., 54:* 141-146.

DUKE, B. O. L., 1964. Studies on loiasis in monkeys. IV. Experimental hybridization of the human and simian strains of *Loa. Ann. trop. Med. Parasit., 58:* 390-408.

DUKE, B. O. L. & WIJERS, D. J. B., 1958. Studies on loiasis in monkeys. I. The relationship between human and simian *Loa* in the rain-forest zone of the British Cameroons. *Ann. trop. Med. Parasit., 52:* 158-175.

GORDON, R. M., KERSHAW, W. E., CREWE, W. & OLDROYD, H., 1950. The problem of loiasis in West Africa with special reference to recent investigations at Kumba in the British Cameroons and at Sapele in Southern Nigeria. *Trans. R. Soc. trop. Med. Hyg., 44:* 11-41.

HADDOW, A. J., 1952. Further observations on the biting habits of Tabanidae in Uganda. *Bull. ent. Res., 42:* 659-674.

DISCUSSION

J. R. Busvine

Is there any reason to suppose that the different infection levels in the three monkey host species for *Loa* are dependent on differences in habits? Or do you suppose that their immune responses differ?

B. O. L. Duke

I suspect that habit differences may come into the picture. Drills, for example, tend to move about in large troops which provide a much greater target and source of attraction to *Chrysops* than do the cercopithecids. I have no information on the immune responses of the different monkey species.

M. T. Gillies

One knows that tabanids are attracted to sources of CO_2. Do you think that it could have been the CO_2 produced by smoky fires which was fortifying the effect of the human catches and so pulling in more flies?

B. O. L. Duke

We tried other sources of CO_2, namely that coming from two large multi-burner Primus stoves and also the smoke from a fire of oily rags and old rope. Neither produced any significant attractive effect on *Chrysops*. We were not able to test the effect of pure CO_2 from a cylinder, but our tentative conclusion was that there was some specific substance in wood smoke which attracted or activated the flies. It is a field which merits further investigation both for *Chrysops* and for other species of man-biting insects.

M. W. Service

Parasites such as *Onchocerca cervicalis* in the horse occur mainly in the belly region, the area where the *Culicoides* vectors prefer to bite. Is there any evidence that microfilariae of *Loa loa* accumulate in the buttocks of the monkey hosts, the region where the *Chrysops* vectors mainly feed?

B. O. L. Duke

This point has not been investigated. In general a localized spatial distribution is a property of skin-dwelling rather than blood-dwelling microfilariae. It is possible that the buttock region may be more richly supplied with capillaries or that the microfilariae accumulate there as a gravitational effect in the sleeping monkey, but we have no direct evidence on this.

A. W. R. McCrae

Has work been done on the relative susceptibility or refractoriness of day-active and night-active *Chrysops* to development of the two periodic types of *Loa*?

B. O. L. Duke

Although no accurate quantitative work has been done on this problem we were able to get good development of either strain of *Loa* in *C. silacea, C. dimidiata, C. langi,* and *C. centurionis.*

S. M. Omer

Did you cross the hybrid parasites back with their parent strains and if so, what sort of periodicities did you get?

B. O. L. Duke

We did experiments in which hybrid males were crossed with virgin females of the two parent strains separately. Details of the periodicities obtained can be found in Duke (1964).

M. G. Taylor

Would it not be possible to regard the simian and human parasites as two sibling species as a result of your hybridization experiments and studies on differences in vector species etc.?

B. O. L. Duke

Yes, I think it would.

D. S. Bertram

Is there evidence for splenic destruction of microfilariae in filarial species other than that in the drill hosts of *Loa*? This control of microfilariae of *Loa* in monkeys by splenic destruction seems unusual.

B. O. L. Duke

I suspect that the drill spleen also destroys microfilariae of the *Dirofilaria* spp. which sometimes co-exist with *Loa* in these monkeys, but I have never seen a pure *Dirofilaria* infection in which the point could be elucidated. The spleen of *Cercopithecus nictitans martini* also destroys *Loa* microfilariae. A similar effect should always be watched for in any monkey infected with any species of microfilaria.

P. Jordan

Was it possible to alter the nocturnal periodicity by altering the sleep cycle of the monkeys?

B. O. L. Duke

As it is possible to do this in man I suspect that it could also be done in monkeys, but we have never been working in conditions where the sleep cycle of infected monkeys could be controlled.

Human behaviour in the transmission of parasitic diseases

G. S. NELSON

London School of Hygiene and Tropical Medicine, London

The effect of human behaviour on the prevalence of helminth infections is discussed in relation to different socio-economic groups with particular reference to infections that man acquires from animals.

The following subjects are used to illustrate the subject:

(1) Trichinosis in Africa among hunter gatherers and detribalized agriculturalists who eat the flesh of the bush pig.

(2) Trichinosis among Eskimoes and explorers in the Arctic who eat the flesh of the polar bear and marine mammals.

(3) Tapeworm infection in nomadic pasturalists.

(4) Hydatid disease resulting from man's intimacy with the dog and the effect of social customs on the prevalence of this disease.

(5) Schistosomiasis in settled agricultural communities.

(6) The menace of keeping pets with special reference to toxocariasis.

(7) The dangers of transmission of new diseases as the result of handling exotic animals.

(8) The dangers of acquiring exotic disease as a result of tourism.

CONTENTS

INTRODUCTION

For someone who is neither an anthropologist nor a behavioural scientist to talk about "Human behaviour in the transmission of parasitic disease" is perhaps foolhardy, especially as I am not even an ecologist but merely a medical naturalist who is interested in the interrelationship of parasitic infections of man and animals. But I have had an opportunity of studying a variety of parasitic diseases in people of diverse cultures in various parts of Africa, and I will therefore try to indicate how human behaviour may affect the distribution and prevalence of a few helminth infections that are particularly subject to social customs.

Many of the more serious manifestations of helminth infections are the result of essentially human activities; the severity of the disease being determined by the intensity of the infection which in turn is often determined by the extent to which man pollutes his environment with his own excreta. It is quite clear that there is very little hope of eliminating the vast amount of helminth disease from the world without drastic changes in human behaviour. But this is the job of the health educationalists. It is an interesting topic but it is not what I want to discuss today and I doubt if you would appreciate a detailed discussion of defecation habits and other unsavoury topics at such an early hour of the morning. Instead I will concentrate mainly on some of the zoonotic diseases that are of interest to the biologist.

The life cycles of the different parasites are so distinct and the behaviour of the people in different regions of the world so diverse that it would be unwise to attempt any form of synthesis or to try to enunciate general principles. It is, however, worth while to look at societies at different stages of development to see to what extent socio-economic and cultural factors affect the pattern of these diseases in the community. To illustrate this I propose to consider some of the helminths that are prevalent amongst the hunter gatherers, the nomadic livestock owners, and settled agriculturalists, and I will finish with a brief comment on the helminth infections in industrial societies. In discussing each group no attempt is made to deal with all the helminth infections that they acquire—to do this would be an immense task—I have merely selected a few examples such as trichinosis, hydatid disease and schistosomiasis to illustrate this neglected aspect of epidemiology.

THE HUNTER GATHERERS

Miscellaneous zoonoses

The only detailed observations of the parasitic fauna of the true hunter gatherers of Africa are those on the bushmen of the Kalahari by Heinz (1961), on the Pygmies of the Congo by Price, Mann, Roels & Merrill (1963) and on the Hadza of Tanzania by Bennett, Kagan, Barnicot & Woodburn (1970). These studies all suggest that the physical environment is perhaps of more importance than human behaviour in determining the pattern of parasitic diseases in the different communities. For example the Pygmies of the rain forest are infected with onchocerciasis but this infection is not seen in the bushmen in the dry Kalahari because the vector is absent from this area. Particular hazards for the hunter gatherers are the zoonoses, that is infections man acquires from wild animals. The highest prevalence of the monkey parasite *Strongyloides fulleborni,* so far recorded in man, is found amongst the Pygmies in the Congo where infection rates may reach more than 50% (Pampiglione, pers. comm.). I also suspect that many of the so-called hookworm infections recorded in the bushman, Hadza and Pygmies are in fact *Ternidens deminutus* the "false hookworm" of monkeys with eggs difficult to distinguish from *N. americanus* and *A. duodenale* (Goldsmid, 1968).

Trichinosis in Africa and the Arctic

Trichinosis which man acquires by eating the undercooked flesh of the wild animal host is likely to be a widespread infection affecting hunter gatherers. There are few records from Africa probably because the disease is difficult to diagnose but it is a serious disease of the hunter gathering Eskimoes in the Arctic regions where there is an entirely feral maintenance cycle in carrion feeding or cannibalistic carnivores with a secondary cycle involving marine mammals and man (Fig. 1). Many species of animals are infected (see Rausch, 1970, for a review of trichinosis in the Arctic). Man intrudes into the zoonotic

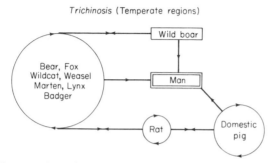

Figure 1. The maintenance of trichinosis in temperate regions.

cycle when he eats the flesh of the polar bear, the black or brown bear or marine mammals such as the walrus and seal. The disease is endemic throughout the Arctic regions and there have been extensive epidemics in Greenland. Transmission can also occur to man through the domestic sledge dog which becomes infected through eating carrion or in extreme conditions by cannibalism when dog eats dog. Fortunately in most parts of the world man has a distaste for the flesh of carnivores including dogs and other carrion feeders—this behavioural characteristic has a considerable effect on the prevalence of the disease in man.

Outbreaks of the disease have occurred amongst sophisticated people who for the sake of adventure or for scientific discovery have intruded into the land of the "hunter gatherer" and who have eaten the flesh of infected animals. It was meat from the polar bear that caused the death of members of the Andree expedition in 1897 when the balloon in which they were attempting to fly over the North Pole came down on an ice floe and the survivors were compelled to live off the flesh of polar bears (Roth, 1950). A very exciting account of this expedition with a semi-fictional reconstruction of the final deaths from trichinosis has been produced by Sundman (1970) in his book *The flight of the Eagle*. I suspect that the severe rheumatism and backache that Nansen suffered during his enforced encampment during the long winter months in Franz Joseph Land where he lived almost entirely off the meat of polar bears and walrus was also due to trichinosis (Nansen, 1897). During the last war there was an outbreak of trichinosis in Spitzbergen in the German garrison (Dege, 1953) and more recently a Russian expedition to the Arctic was seriously affected by trichinosis from eating polar bear meat (Ozeretskovskaya & Upensky, 1957). The main problem amongst the indigenous Eskimoes and the explorers has been a shortage of fuel to adequately cook the meat.

My own experience with trichinosis has been mainly in Africa where the situation is remarkably similar to the Arctic. Here there is a maintenance cycle in the large carrion feeding and cannibalistic carnivores such as the hyaena and jackals with a secondary cycle involving the bush pig and warthog (Fig. 2). Man rarely eats flesh of the primary maintenance hosts in Africa except for ritualistic purposes when they are totem animals as for example the lion or leopard. Most human infections have come from eating the flesh of the bush pig or warthog. The first outbreak recorded in tropical Africa was in Kenya where 11 youths who had been hunting on the slopes of Mt Kenya were admitted to hospital with the heaviest trichinosis infection ever recorded in man (Forrester, Nelson & Sander, 1961). To begin with it was suspected that the infection must have been introduced into the country, but enquiries

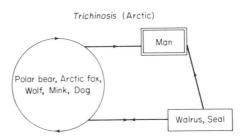

Figure 2. The maintenance of trichinosis in the Arctic.

showed that all 11 patients had eaten undercooked meat of a wild bush pig which had been killed in the forest on the upper slopes of Mount Kenya, and subsequent studies have brought to light almost identical outbreaks in other parts of the country. Final proof of the indigenous enzootic was obtained by examining nearly 2000 mammals representing 50 species. The parasite was found in the bush pig, the leopard, the jackal, the serval, the lion, the spotted and striped hyaenas and in the domestic dog, but never in the domestic pig nor in domiciliated and wild rodents, nor in the numerous viverids and mustelids which feed mainly on rodents (Nelson, Guggisberg & Mukundi, 1963).

The outbreak was new to Kenya because of a change in human behaviour. It was a traditional Kikuyu custom to disdain the flesh of wild animals but as a result of the Mau Mau insurrection many of the old traditions were abandoned and the "teenage" Kenyans with no respect for their elders ate the undercooked meat of the bush pig with dire consequences. In other parts of Africa the flesh of the bush pig and warthog is greatly relished except by Moslems and trichinosis must be widespread where the hunters cook the meat over an open fire toasting only the outside and leaving the centre fully infective. In the Kenya outbreak no women or small children were infected because they ate only well-cooked meat.

Again in Africa as in the Arctic sophisticated immigrants are at risk if they eat the flesh of wild animals and an outbreak of the disease occurred in Dakar following a feast in which the main course was a barbecued warthog (Gretillat & Vassilades, 1967). The strain of parasites isolated from the wild carnivores of Africa and the Arctic are unusual in that they are of exceptionally low infectivity to rats and domestic pigs so that there is no domestic cycle in these areas (Nelson, Blackie & Mukundi, 1966; Nelson, 1970).

THE NOMADIC LIVESTOCK OWNERS

My experience with the pastoralists has been with the Masai and Turkana in Kenya and in referring to these two communities I have selected cestodes as the main topic for discussion

Taeniasis

Taenia saginata is practically universal amongst the cattle owners in East Africa and here we have what Garnham (1958) called the perfect zoonosis or Euzoonosis where there is an obligatory association between man and cattle for the perpetuation of the organism in nature. We examined a wide variety of carnivores in a search for an alternative definitive host of *T. saginata* in Kenya and a large number of antelopes for alternative intermediate hosts but apart from the presence of cysts in a Gnu we drew a blank (Nelson, Pester & Rickman, 1965). The only part of the world where an alternative host may be of significance is in the U.S.S.R. where Krotov (1961) found that reindeer were infected.

The traditional diet of the Masai and Turkana is blood and milk but sufficient undercooked beef is eaten to maintain a high infection rate of *T. saginata* in man. The Manyatta encampments which are surrounded with thornscrub fences to protect the cattle from lions are ideally designed for a high level of transmission. The tapeworm proglottids escape from the anus of the semi-naked warriors as they stand around the herds and the cattle which are often salt hungry will lick up the proglottids from the ground. This ensures that they have perhaps the highest cysticercosis rate anywhere in the world. Fortunately man rarely develops the cysticercosis stage of *T. saginata* but cysts of the pork tapeworm *T. solium* and possibly other species of *Taenia* that are widespread in carnivores are probably responsible for a good deal of epilepsy and other neurological diseases in Africa. It is here that we have an unusual example of human behaviour determining the prevalence of a serious disease. In South Africa there is an increased prevalence of cysticercosis due to the good intentions of the witch doctors in treating tapeworms. Heinz & MacNab (1965) have shown that cysticercosis can result from the ingestion of a concoction of brewed tapeworms which is incorparated in a medicament brewed by the witch doctors. The practice of using a "hair of the dog" to cure disease is not restricted to witch doctors in Africa. Helminths and animal faeces have often been used in preparing medicaments in Europe and Asia. Hoeppli (1959) in his *Parasites and parasitic infections in early medicine and science,* quotes several entertaining examples. Apparently Martin Luther (1483-1546) was treated with a mixture of garlic and horse dung and made no strong objections; fortunately the parasites of the horse are fairly host specific or we might have missed the Reformation. *Ascaris lumbricoides* has been a favourite with pharmacists and there seems little doubt that a ground-up concoction of this parasite had anthelmintic properties; the violent allergic reaction that occurs in people who are hypersensitive to *Ascaris* would make it a potent remedy but it probably also precipitated attacks of asthma.

Another tapeworm that is found in the Masai is the plerocercoid of a species of *Diphyllobothrium* which produces sparganosis, first named by Sambon

(1907) *Sparganum baxteri* after the medical officer who recovered a specimen from an abscess on the leg of a Masai. The adult parasite is a Spirometrid tapeworm which is widespread in hyaenas (Nelson *et al.,* 1965; Sachs, 1969; Muller, Weeks & Nelson, 1968). The Masai have no burial custom and they leave their dead for the hyaenas to eat so in this way man can be the intermediate host of a tapeworm of wild carnivores. This is not an entirely unique situation because in areas where there is the same custom for disposal of one's relatives the carnivores can become infected with *Trichinella* and *Echinococcus* of human origin.

Hydatid disease

The great curse of the pastoralist is the dog that he uses to help herd his flocks. It allows him to take life in a leisurely manner but at the same time it is a constant threat to himself, his family, and his livestock because it is the dog that is the host *par excellence* of *Echinococcus.* Fortunately hydatid disease is not universal amongst the dog-owning nomadic pastoralists. It may be present in one tribe and absent in the next. For example in Kenya it is rarely seen in the Masai and yet one of the highest infection rates in the world is in the Turkana. This interesting focus has been studied by Wray (1958), Nelson & Rausch (1963) and by Schwabe (1969) who discusses the ecological and epidemiological aspects of this problem in his scholarly work *Veterinary medicine and human health.* The prevalence of hydatid disease in man is determined to a very large extent by man's behaviour; wherever a high prevalence rate is found there are usually sociological factors that account for this. The most important consideration is the level of intimacy of the people with infected dogs. The difference in prevalence rates between the Masai and Turkana is probably because the Masai have no great love of their dogs and keep them at a distance whereas the Turkana have a tradition of intimate contact at a very early age. In parts of Turkana a "dog-nurse" is used to eat the excreta of babies and clean up the mess when the child vomits or defaecates. In this way the dog which has infective eggs around its muzzle (due to licking its own anus) will transfer the infection to the child at the most susceptible age for developing a hydatid cyst. Although the neighbouring Suk tribe live in the same type of environment and also use dogs to herd their sheep and cattle this custom is absent and hydatid disease is rare.

The high prevalence rate in Iceland at the end of the last century was due to the very intimate contact with infected dogs in the houses during the long dark Arctic winters. The disease was successfully controlled in Iceland when as a result of improved education the people understood the life cycle of the parasite and completely changed their attitude to the dogs and banished them from the houses. The eggs of *Echinococcus* are extremely resistant to physical conditions and remain infective for long periods; this probably explains the unusually high prevalence amongst shoemakers in the Lebanon who at one time used dog faeces for tanning leather (Schwabe & Daoud, 1961). It is a common sight to see a shoemaker soften the sole of a shoe with saliva to make the leather more pliable. Other occupational groups that are vulnerable to hydatid disease are butchers and abattoir workers who have access to condemned meat containing hydatid cysts. Quite unwittingly they feed the infected meat to their dogs and as a result expose their families to an increased risk of infection.

Religion can also affect the prevalence of hydatid disease. Mohamed wisely forbad the close association of man with dogs and is reputed to have said that "Angels do not enter a house where there is a dog"; the result is that in countries like the Lebanon there is a significantly higher incidence in Christians as compared to Moslems (Schwabe, 1969).

The custom of feeding foxhounds on infected horse meat from the knackeries ensures an especially high prevalence of *Echinococcus* in foxhounds and horses in England but surprisingly the disease is not an occupational hazard for the kennel maids and huntsmen who look after the hounds. It is possible that this particular strain of the parasite is of low infectivity for man. Several "strains" with differences in their infectivity to the intermediate and definitive host have been recorded (Williams & Sweatman, 1963). The parasite in dogs, foxes and horses has been named *E. granulosus equinus* and it is the only form found in Ireland where there is no indigenous transmission of hydatid disease to man; an observation which emphasizes that in some instances it is the infectivity of the parasite rather than human behaviour that determines the prevalence of parasitic diseases.

THE SETTLED AGRICULTURALISTS

Schistosomiasis

In this group which represents most of mankind the social customs and the patterns of behaviour, relevant to the transmission of helminth infections, are so diverse that I propose to deal only with one disease, namely schistosomiasis. With this disease we have the paradox that in the endemic areas any increase in the economic development of the community is likely to lead to an increase in the prevalence of the disease. This is certainly true of schistosomiasis in Africa and South America. On the other hand there is evidence that *S. japonicum* is being successfully controlled in China and Japan mainly due to changes in agricultural practices and in human behaviour.

In the Far East schistosomiasis is a zoonosis with a wide variety of wild and domestic animal hosts. At one time the cow was considered to be an important source of infection for man in Japan but according to Yokogawa (1970) the substitution first of horses and then of tractors for ploughing has been a major factor in reducing transmission. At the same time the intermediate snail host has been reduced by clearing the canals and lining them with cement. Another factor that has affected transmission has been the strict supervision of methods for preparing human faeces as a fertilizer. In China the practice of washing chamber pots in the canals has been forbidden and this is thought to be one of the changes that has reduced the infection rates in snails (Tien-Hsi Cheng, 1971). The ideal method of preventing this disease is to keep people away from infected water. But who can be so heartless as to prevent children from swimming and washing in the only available water in countries where the midday temperature is usually over 90° and how can you keep a rice farmer out of his paddy fields? In South Africa Pitchford (1966) has shown that it is only when alternative water supplies are made available that this method of control is successful.

Unfortunately the prospects for the control of schistosomiasis throughout most of Africa and South America are much less promising than in Asia mainly because developments for irrigation and for the production of hydro-electricity are proceeding too rapidly for the public health authorities to be able to cope. The Aswan dam in Egypt is a frightening example of such a development in a country where control has been largely ineffective in spite of a great deal of effort over many years. Before the dam was built there was a restricted breeding season for the intermediate snail hosts and *Biomphalaria*, the intermediate host of *S. mansoni*, was confined to the Delta region but water is now available for perennial irrigation in areas where before there was only a seasonal irrigation from the annual Nile floods and these snails have extended their distribution up the Nile valley for several hundred miles. This will soon be followed by an extension of the disease. There is also a strong possibility that the disease will be worse because of the prolongation of the transmission season and the consequent increase in intensity of infection.

The Volta Dam in Ghana is another example of a major economic development resulting in the spread of schistosomiasis. Many of the people who have settled in new villages along the edge of this man-made lake, which is now one of the largest in the world, are already heavily infected with *S. haematobium*.

The difficulties of preventing the spread of the disease in irrigated areas is well illustrated by the Gezira scheme in the Sudan. Here in what was formerly part of the arid Sahara more than 2,000,000 acres are under irrigation from a single water source on the Sennar dam. From the very beginning efforts were made to prevent the spread of the disease but again control has been ineffective in spite of the efforts of a devoted staff and a very considerable expenditure. Perhaps too much attention has been given to snail control and to chemotherapy and not sufficient to studying human behaviour. Apart from the work of Husting (1968) in Rhodesia and the more recent study by Jordan and his colleagues in St Lucia there have been no deliberate scientific observations on defaecation habits and water contact in relation to the transmission of schistosomiasis. The observations by Bennett *et al.* (1970) which show that the primitive Hadza tribe in Tanzania are practically free of schistosomiasis whereas the neighbouring Bantu tribes in Sukumuland are heavily infected is of interest. It is suggested that the low incidence amongst the Hadza is because they rarely wash or play in water and that they collect their drinking water from holes dug in dry river beds. Hamadryas baboons in Ethiopia apparently use a similar technique and drink filtered water from holes at the sides of streams—a method which Altman & Altman (1970) suggest might have evolutionary value in protecting them from schistosomiasis. Further south in East Africa baboons are naturally infected with *S. mansoni* but the intensity of infection is usually light, possibly because baboons very rarely swim in infected water. They may become infected by eating infected snails. This is one potential source of protein that man avoids; although he greatly relishes marine and land molluscs he rarely eats the abundant aquatic snail hosts of schistosomes and other trematodes.

The custom of eating land snails is, however, not without danger. The dissemination of the giant African land snail *Achatina* in Asia has resulted in unforeseen epidemics of eosinophilic meningitis in man in the Pacific. This snail which was deliberately introduced to many of the Pacific islands by the Japanese has proved to be an excellent intermediate host of the rat parasite *Angiostrongylus cantonensis*. When infected snails are eaten by man the nematode larvae migrate to the central nervous system and produce a serious clinical syndrome (Alicata, 1966). This unexpected outbreak should be taken as a lesson by those who are advocating the biological control of schistosomiasis by dissemination of the large South American snail *Marisa* throughout Africa. We have no idea what the creature will do in its new environment. At a recent conference dealing with the problem of biological control of schistosomiasis in Africa an enthusiastic American exponent of control by *Marisa* was asked by a delegate from one of the African countries if *Marisa* ate not only snails but also young rice. He was told that *Marisa* does indeed eat very young rice seedlings but not when they have grown to the size when they are being transplanted in the fields. The delegate from the small African country was not entirely satisfied with the answer and said that he was not sure how primitive the agriculturalists were in the United States but in his country rice seeds were planted from aeroplanes and he thought *Marisa* would destroy the rice crops!

Another example of an unforeseen ecological hazard resulting from human behaviour has been the increase in the baboon population and the consequent increase in the reservoir of schistosomiasis in East Africa as a result of their protection in game reserves and also as the result of the vanity of ladies in affluent societies who wear leopard skin coats, the leopard being the main predator of the baboon.

THE INDUSTRIAL SOCIETY

Trichinosis

I have selected trichinosis for further discussion in this section because until quite recently trichinosis was more prevalent in the advanced societies than anywhere else. The highest incidence was in the United States of America where the consumption of pork is the highest in the world. Why we of the Western World are so addicted to eating pork is something of a mystery especially as we are constantly reminded of the dangers when we read the Bible. Moses in Deuteronomy, Chapter 14 says "and the swine it is unclean unto you—Ye shall not eat of their flesh nor touch their dead carcasses—Ye shall not eat anything that dieth of itself". It is likely that some of our forefathers were sceptical of accepting this particular Law of Moses because he was rather an unethical fellow. In referring to the meat from dead animals he says, "Thou shalt give it unto the stranger that is in thy gate that he may eat it and thou mayest sell it unto an alien"! Strict Jews and also the followers of Mohamed have, however, persisted in their distaste of pigs and trichinosis is completely absent from these communities.

The high level of transmission in Christian Europe and America at the beginning of the century was due to industrialized man's extraordinary extravagance and his disgusting behaviour in making pigs cannibals by feeding them on garbage containing infected pork products. Fortunately garbage not only spread trichinosis but also vesicular exanthema, a virus infection of swine, and legislation was introduced to prevent the feeding of uncooked garbage to swine, not as a measure against the human disease but against the virus disease that was devastating the pig population in America. As a by-product trichinosis transmission has been greatly reduced and vesicular exanthema has been eliminated (one of the few examples of the total extermination of a pathogenic organism). There is still some trichinosis transmission amongst pigs and it has been suggested that the light infections that are usually recorded are the result of pigs eating one another's tails in their overcrowded sties (Zimmerman, Hubbard, Schwarte & Biester, 1962). Occasionally they eat an infected rat but rats like pigs are probably not true maintenance hosts. They acquire their infection on dumps and in sewers where they have access to infected pork refuse (see Fig. 3).

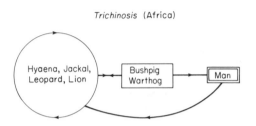

Figure 3. The maintenance of trichinosis in Africa.

The commonest source of infection in Britain is the undercooked sausage and women are far more often infected than men. This is not surprising as many women will eat raw sausages or at least have a nibble at the sausage meat when preparing a meal; their husbands and children only suffer if the food is improperly cooked. In a recent outbreak in Ireland it was found that 55 out of a total of 76 cases admitted to eating raw sausages (Corridan & Gray, 1969). Because of the possible effect on exports in Ireland the authorities have been reluctant to admit that trichinosis is endemic and they have argued that it is an imported disease. Such behaviour in the face of overwhelming evidence to the contrary might well have resulted in further epidemics. The authorities have now been convinced that the parasite is enzootic in Ireland as a result of the further demonstration by Corridan, O'Rourke & Verling (1969) of *T. spiralis* in red foxes.

One of the most extensive outbreaks of trichinosis ever recorded occurred a few years ago in Poland and again the authorities of this pork producing country have been reluctant to publish the details. There were several hundred cases due to the widespread distribution of a particular type of sausage that is traditionally eaten uncooked on a particular feast day. The sausages were heavily infected and had been prepared with meat from pigs that had been fed on the carcasses of silver foxes that had been reared for their skins. In Poland there is a strict system of inspection of carcasses but it is rumoured that on this occasion there was a dereliction of duty due to too convivial a consumption of vodka by the meat inspector and the pig farmer!

One of the most unusual outbreaks of trichinosis was reported by Rubli from Switzerland (1936) where the source of infection was the improperly cooked flesh of the large South American rodent *Myocastor coypu* that was being reared for its nutria fur. There are obvious dangers in such exotic tastes. The present craze in expensive restaurants of serving up dishes prepared from the flesh of wild animals can also be dangerous as was only too obvious when a group of six gourmets from New York developed severe trichinosis after eating a variety of dishes made from the flesh of a black bear (Roselle, Schwatts & Geer, 1965). The modern craze for barbecues can be a similar menace especially in countries with poor standards of meat hygiene. Anyone eating an undercooked steak in most parts of Africa has a very good chance of developing a *Taenia* infection. One of the most expensive of all foods, caviare, is not without its dangers and there have been cases of diphyllobothriasis developing in people addicted to this delicacy.

Toxocariasis

Although industrialized man is insulated from most of the zoonoses that are maintained in nature by wild animals he is at considerable risk from infections maintained by livestock and pets. Jenner was probably the first British scientist to recognize the danger of keeping pets. In his famous enquiry into the cause and effect of cowpox published in 1798 he says, "The deviation of man from the state in which he was placed by nature seems to have proven to him a prolific source of diseases. From the love of splendour, from the indulgence of luxury and from his fondness for amusement he has familiarized himself with the great number of animals which may not usually have been intended for his amusement. The wolf disarmed of his ferocity is now pillowed in the Lady's lap. The cat, the little tiger of our island, whose natural home is the forest is equally domesticated and caressed." This was a prophetic statement as at the time Jenner knew nothing about the life cycles and transmission of the many parasitic infections that man shares with his pets. A few years ago I drew up a list of the pathogens that had been recorded as being shared by man and dogs and found that there were more than 50 different organisms that might be transmitted either from dog to man or vice versa. One of the last organisms to be added to this list was *Toxocara*. As a result of the studies by Beaver and his colleagues in America (Beaver, Snyder, Carrera, Dent & Lafferty, 1952; Beaver, 1956, 1959) and Woodruff and his colleagues in this country (see Woodruff, 1970, for a recent review) we now know that toxocariasis is responsible for a considerable amount of serious disease especially in children. Several parasites of cats and dogs are involved but the most important species is *T. canis* which is transmitted transplacentally from the bitch to her pups. This is a cosmopolitan parasite. The infection rate in dogs in England is very high and the eggs are widely disseminated in the environment. If a child ingests the eggs from the puppies' faeces he will develop an abortive infection; if this is severe the migrating larvae will cause enlargement of the liver, an irregular temperature and eosinophilia, they may invade the eye and produce retinal granulomas which can cause blindness or be mistaken for tumours with the resulting unnecessary enucleation of the eye. The parasitic aetiology is rarely recognized except by retrospective study of serial sections (Wilder, 1950; Ashton, 1969). The disgusting state of the pavements and public parks in many of our cities where urban dogs are allowed by their owners to spread the infection makes one wonder why toxocariasis is not more prevalent.

Exotic disease

We are fortunate in the United Kingdom in that strict quarantine regulations can be enforced to prevent the importation of animals with the more obvious diseases transmissible to man. So far we have been successful in preventing the indigenous transmission of rabies in our wildlife and pets but some of the infections of animals cannot be detected. The frightening outbreak of fatal virus infections in Germany due to handling monkeys in a laboratory preparing poliomyelitis vaccine and the transmission from patients to doctor outside the laboratory is a sufficient warning of the dangers of handling exotic animals (see Martini, 1969; Simpson, 1969; Zlotnik, 1969).

If we are vigilant we can prevent most of the diseases that are likely to be introduced by animals but there is no effective quarantine against man importing exotic diseases himself. One can predict that one of the greatest hazards to the more highly developed societies will come as a result of man's "love of splendour from the indulgence of luxury and from his fondness for amusement". Every year we see more and more tourists searching for the sun and adventure in areas which a few years ago were regarded as "the white man's grave"; they live in a fool's paradise unaware that very little progress has been made in the control of parasitic and other infectious diseases in the areas they are visiting. Within the next few years we will see vast hordes of sun loving tourists on safari in Africa, Asia and South America bathing on the coral beaches, dancing under moonlit skies, yachting on the inland lakes and eating the exotic foods quite oblivious of the hookworm larvae in the sand, the schistosomes in the lakes, the *Ascaris* eggs and cholera vibrios on the salads and the host of virulent organisms in the biting insects which they regard as no more than a nuisance. Usually these will be "dead end" infections with no further transmission in the non-tropical areas. But we cannot always be sure. At the present time there is the threat of a major extension of malaria, not transmitted by mosquitoes, but syringe-transmitted by drug addicts returning from South East Asia. This is no mere fantasy but it is already a reality in parts of the United States where drug addicted soldiers returning from Vietnam have transmitted malaria to people who have never visited an endemic area.

REFERENCES

ALTMAN, S. A. & ALTMAN, J., 1970. *Baboon ecology. African field researcher:* 220. Chicago: University of Chicago Press.

ALICATA, J. E., 1966. The presence of *Angiostrongylus cantonensis* in islands of the Indian Ocean and probable rôle of the giant African snail, *Achatina fulica* in dispersal of the parasites to the Pacific islands. *Can. J. Zool., 44:* 1041-1049.

ASHTON, N., 1969. *Toxocara canis* and the eye. In P. V. Rycroft (Ed.), *Proc. 2nd Int. Corneo-plastic Surgery Conf.:* 579-591. Oxford: Permagon Press.

BEAVER, P., 1956. Larva migrans. A review. *Expl Parasit., 5:* 587-621.

BEAVER, P. C., 1959. Visceral and cutaneous larva migrans. *Publ. Hlth Rep. Wash., 74:* 328-332.

BEAVER, P., SNYDER, H., CARRERA, G., DENT, J. & LAFFERTY, J., 1952. Chronic eosinophilia due to visceral larva migrans; report of three cases. *Pediatrics, 9:* 7-19.

BENNETT, F. J., KAGAN, I. G., BARNICOT, N. A. & WOODBURN, J. C., 1970. Helminth and protozoal parasites of the Hadza of Tanzania. *Trans. R. Soc. trop. Med. Hyg., 64:* 857-880.

CORRIDAN, J. P. & GRAY, J. J., 1969. Trichinosis in South-West Ireland. *Br. med. J., 2:* 727-730.

CORRIDAN, F. J., O'ROURKE, F. J. & VERLING, M., 1969. *Trichinella* spiralis in the red fox (*Vulpes vulpes*) in Ireland. *Nature, Lond., 222*(5166): 1191.

DEGE, W., 1953. Trichinen-Infektionen in der Arktis und S.A. Andrees Tod. *Naturw. Rdsch., Stuttg., 6:* 332.

FORRESTER, A. T. T., NELSON, G. S. & SANDER, G., 1961. The first record of an outbreak of trichinosis in Africa south of the Sahara. *Trans. R. Soc. trop. Med. Hyg., 55:* 503-513.

GARNHAM, P. C. C., 1958. Zoonoses or infections common to man and animals. *J. trop. Med. Hyg., 61:* 92-94.

GOLDSMID, J. M., 1968. The differentiation of *Ternidens deminutus* and hookworm ova in human faeces. *Trans. R. Soc. trop. Med. Hyg., 62:* 109-116.

GRETILLAT, S. & VASSILADES, G., 1967. Présence de *Trichinella spiralis* (Owen, 1835) chez les carnovires et suidés sauvages de la region du delta du fleuve Sénégal. *C. r. hebd. Séanc. Acad. Sci., Paris, 264:* 1297-1300.

HEINZ, H., 1961. Factors governing the survival of bushmen worm parasites in the Kalahari. *S. Afr. J. Sci., 57:* 207-213.

HEINZ, H. J. & McNAB, G. M., 1965. Cysticercosis in the Bantu of Southern Africa. *S. Afr. J. med. Sci., 30:* 19-31.

HOEPPLI, R., 1959. *Parasites and parasitic infections in early medicine and science,* 526 pp. Singapore: University of Malaya Press.

HUSTING, E. L., 1968. *A biological and sociological study of the epidemiology of bilharziasis.* University of London Ph.D. Thesis.

JENNER, E., 1798. *An enquiry into the causes and effects of the variolae vaccinae, a disease discovered in some of the western counties of England, particularly Gloucestershire and known by the name of cow pox.* London: Printed for the author by Sampson Low, 1798.

KROTOV, A. I., 1961. The epidemiology of *Taenia saginata. Medskaya Parazit., 30:* 98-99.

MARTINI, G. A., 1969. Marburg agent disease: in man. *Trans. R. Soc. trop. Med. Hyg., 63:* 295-302.

MULLER, R. L., WEEKS, B. & NELSON, G. S., 1968' The laboratory maintenance of a cestode suspected of causing sparganosis in East Africa. *Trans. R. Soc. trop. Med. Hyg., 62:* 467.

NANSEN, F., 1897. *Farthest north, II:* 729. New York: Harper.

NELSON, G. S., 1970. Trichinosis in Africa. In S. E. Gould (Ed.), *Trichinosis in man and animals:* 473-493. Springfield, Illinois: Charles. C. Thomas.

NELSON, G. S. & RAUSCH, R. L., 1963. *Echinococcus* infections in man and animals in Kenya. *Ann. trop. Med. Parasit., 57:* 137-149.

NELSON, G. S., GUGGISBERG, C. W. A & MUKUNDI, J., 1963. Animal hosts of *Trichinella spiralis* in East Africa. *Ann. trop. Med. Parasit., 57:* 332-346.

NELSON, G. S., PESTER, F. R. N. & RICKMAN, R., 1965. The significance of wild animals in the transmission of cestodes of medical importance in Kenya. *Trans. R. Soc. trop. Med. Hyg., 59:* 507-524.

NELSON, G. S., BLACKIE, E. J. & MUKUNDI, J., 1966. Comparative studies on geographical strains of *Trichinella spiralis. Trans. R. Soc. trop. Med. Hyg., 60:* 471-480.

OZERETSKOVSKAYA, N. N. & UPENSKY, S. M., 1957. Group infection by trichinellosis from the meat of white bear in the Soviet Artic Region. *Medskaya Parazit., 26:* 152-159.

PITCHFORD, R. J., 1966. Findings in relation to schistosome transmission in the field following the introduction of various control measures. *S. Afr. Med. J. (Suppl.), 40:* 3-16.

PRICE, D. L., MANN, G. V., ROELS, O. S. & MERRILL, J. M., 1963. Parasitism in Congo pigmies. *Am. J. trop. Med. Hyg., 12.* 383-387.

RAUSCH, R. L., 1970. Trichinosis in the Artic. In S. E. Gould (Ed.), *Trichinosis in man and animals:* 348-374. Springfield, Illinois: Charles C. Thomas.

ROSELLE, H. A., SCHWARTZ, D. T. & GEER, F. G., 1965. Trichinosis from New England bear meat: Report on an epidemic. *New Eng. J. Med., 272:* 304-305.

ROTH, H., 1950. Nouvelles expériences sur la trichinose avec considerations speciales sur son existence dans les regions arctiques. *Bull. Off. int. Epizoot., 34:* 197-220.

RUBLI, H., 1936. Trichinose beim Sumpfbiber *Myocaster coypus. Schweizer Arch. Tierheilk., 78:* 420-424.

SACHS, R., 1969. Cysticercosis, echinococcosis and sparganosis in wild herbivores in East Africa. *Vet. med. Rev., 2:* 104-114.

SAMBON, L. W., 1907. *Sparganum baxteri* in connective tissue of man. *Proc. zool. Soc., Lond.,* 282-283.

SCHWABE, C. W. & ABOU DAOUD, K., 1961. Epidemiology of echinococcosis in the Middle East. I. Human infections in the Lebanon. 1949-1959. *Am. J. trop. Med., 10:* 375-381.

SCHWABE, C. W., 1969. *Veterinary medicine and human health:* 713 pp. London: Baillière, Tindall & Cassell.

SIMPSON, D. I. H., 1969. Marburg agent disease: in monkeys. *Trans. R. Soc. trop. Med. Hyg., 63:* 303-309.

SUNDMAN, P. O., 1970. *The flight of the eagle:* 383pp. (Transl. by M. Sandbach). London: Secker & Warburg.

TIEN-HSI CHENG, 1971. Schistosomiasis in mainland China. A review of research and control programs since 1949. *Am. J. trop. Med. Hyg., 20:* 26-53.

WILDER, H. C., 1950. Nematode endophthalmitis. *Trans. Am. Acad. Ophthal. Oto-lar., 55:* 99.

WILLIAMS, R. J. & SWEATMAN, G. K., 1963. On the transmission, biology and morphology of *Echinococcus granulosus equinus*, a new sub-species of hydatid tapeworm in horses in Great Britain. *Parasitology, 53:* 391-408.

WOODRUFF, A. W., 1970. Toxocariasis. *Br. med. J., 3:* 663-669.

WRAY, J. R., 1958. Note on human hydatid disease in Kenya. *E. Afr. med. J., 35:* 37-39.

YOKOGAWA, M., 1970. Schistosomiasis in Japan. In Manabu Sasa (Ed.), *Recent advances in research on filariasis and schistosomiasis in Japan:* 231-255. Manchester, England. Publ. University of Tokyo Press, Tokyo Park Press.

ZIMMERMANN, W. J., HUBBARD, E. D., SCHWARTE, L. H. & BIESTER, H. E., 1962. Trichinosis in Iowa swine with further studies on modes of transmission. *Cornell Vet., 52:* 156-163.

ZLOTNIK, I., 1969. Marburg agent disease: pathology. *Trans. R. Soc. trop. Med. Hyg., 63:* 310-323.

DISCUSSION

J. E. D. Keeling

I have recently discovered a mixed *Ascaris/Trichuris* infection in a two-year-old child living in a fashionable area of Surrey. The child has never left that area and has not yet started school. Would Professor Nelson care to comment on the occurrence and transmission of helminthic infections in that special centre of sophistication, the Home Counties?

G. S. Nelson

Indigenous infections with *Trichuris* still occur from time to time in England especially in hospitals dealing with mentally retarded children but *Ascaris* transmission is surprisingly rare. I have often wondered why we don't see more cases of indigenous infection of both parasites. We eat plenty of imported fruits and vegetables from countries where these helminths are prevalent. In the case referred to by Dr Keeling it might be worth looking at the faeces of the parents or close relatives, even the "au pair" girl if they have one. As I have said in the paper, tourism is likely to be a fruitful source of exotic parasitic infections.

D. Davenport

What relationship to parasitology in particular and the evolution of human symbioses in general, has the vast increase in the number of uncared-for dogs in the United States?

G. S. Nelson

"Man's best friend" shares with his master more than 50 different organisms belonging to all the main groups of pathogens but fortunately in only a few instances is the dog an obligatory maintenance host of the parasite. In the past we have been mainly concerned with rabies and hydatid disease but toxocariasis is being increasingly recognized as an aetiological agent in the wide spectrum of human disease. I have been appalled by the filthy state of the pavements in New York which are fouled not only by "un-cared" dogs but by expensive poodles on their way to be manicured by the local hair stylists. It might help to make American cities and those in this country pleasanter places to live in if the dog were to evolve a really noxious parasite. This might take it out of the lady's lap and return it to the open field where it belongs.

G. C. Coles

Could you comment on the control of schistosomiasis by altering human behaviour patterns?

G. S. Nelson

We are likely to have even less success in preventing children in the tropics from bathing in infected water or preventing indiscriminate defaecation than the health educationalists in the western world have had in preventing people from committing suicide by smoking.

Modification of intermediate host behaviour by parasites

JOHN C. HOLMES AND WILLIAM M. BETHEL

University of Alberta, Edmonton, Alberta, Canada

Recent concepts of predator-prey relationships, particularly concerning characteristics determining which prey individuals are taken, are reviewed. This review leads to the conclusion that parasites may adopt different evolutionary strategies in altering the behaviour of their intermediate hosts to increase the vulnerability of the latter to predation by the definitive hosts. These strategies include reducing the stamina, increasing the conspicuousness, disorienting, and altering the responses of the intermediate hosts. Evidence that parasites have adopted these strategies is reviewed.

Evidence is presented to show that cystacanths of the acanthocephalan *Polymorphus paradoxus* reverse the phototaxis and produce an altered evasive response in *Gammarus lacustris,* and that these altered responses increase the vulnerability of infected gammarids to mallard ducks.

CONTENTS

INTRODUCTION

Parasites are marvellously well-adapted organisms. Many of their most interesting adaptations involve intimate, interrelationships with their hosts. Such interrelationships may involve any, or all, aspects of the biology of the parasites, including their morphology, physiology, immunochemistry, or (as evidenced by this symposium) their behaviour. In many of these interrelationships parasites have co-opted the responses of their hosts to provide not only the environment necessary for the parasite's growth and reproduction, but also the feed-back loops necessary for regulation of the parasite's population within the host (Ractliffe, Taylor, Whitlock & Lynn, 1969; Kennedy, 1970), or even factors inhibiting the parasite's competitors (Schad, 1966). It is not surprising, therefore, that parasites have used the behavioural patterns of their hosts to assure transmission. In some cases this has involved modification of the normal behaviour of infected intermediate hosts.

A few authors have shown modified behavioural patterns in snails infected with sporocysts or rediae. Rothschild (1940) found numerous snails infected with *Cercaria pricei* Rothschild on top of vegetation at the surface of a pond, whereas all uninfected snails were below the vegetation, suggesting that the normal diurnal migration to the deeper water was hindered by the trematode larvae. Sindermann (1960) and Sindermann & Farrin (1962) also postulated reduced mobility or other interference with the usual seasonal migration to explain concentrations of marine snails infected with *Austrobilharzia vari-glandis* (Miller & Northrup) and *Cryptocotyle lingua* (Creplin) in the high tide zones in the fall and winter. In the last two cases, the authors suggested that the altered behaviour of the infected snails increased the probability of the cercariae infecting the next host. Lambert & Farley (1968) substantiated Sindermann & Farrin's observations on *Littorina littorea* (L.) infected with *C. lingua.* Using a tide-simulating machine, Lambert & Farley were able to show that infected snails reacted more slowly to the cold atmosphere and migrated to a lesser extent, but that the basic response was the same as that of uninfected snails. Similarly, Etges (1963) showed that *Austrolorbis glabrata* (Say) infected with *Schistosoma mansoni* Sambon were less sensitive to a chemical attractant (wheat germ), but reacted similarly to uninfected snails. The significance of the decreased sensitivity was not investigated.

Carney (1969) reviewed the effects of metacercariae of the liver fluke *Dicrocoelium dendriticum* (Rudolphi) on the behaviour of their second intermediate hosts, formicine ants. Hohorst and his co-workers (Hohorst & Graefe, 1961; Hohorst & Lämmler, 1962; Hohorst, 1964) showed that most of the cercariae ingested by the ant penetrated into the abdominal haemocoele and encysted, but that one (or occasionally two or three) migrated to the suboesophageal ganglion, where it encysted close to the nerves to the mouthparts. These "Hirnwürmer", or brainworms, were not infective to definitive hosts, but were associated with a marked change in the behaviour of the infected ants. Infected ants were attached to the tops of blades of grass or leaves of plants during the cooler parts of the day, when uninfected ants had returned to the nest. Anokhin (1966) and Grus (1966) showed that the behavioural change was temperature-dependent, and possibly also light- or humidity-dependent, with the ants migrating up the vegetation and grasping the plants with their mandibles as the temperature decreased in early evening, remaining torpid in such locations throughout the night, and becoming reactivated with rising temperatures during the morning. At midday, their activity is apparently similar to that of uninfected ants. This behavioural pattern keeps the infected ants near the top of the vegetation during the early morning and late evening grazing periods of their ungulate definitive hosts, but allows them to move to less exposed areas during the hot, dry midday. This is an elegant example of the case in which infective larvae have modified the behaviour of the host to increase the probability of being ingested *accidentally* by the definitive host.

Another interesting case, and one amenable to laboratory investigation, is that of an infective stage of a parasite which modifies its intermediate host's behaviour in such a way as to increase the probability of *intentional* predation by the definitive host. This case is the subject of the balance of our paper.

Wright, C. A. (1966), in a review of the pathogenicity of helminths in molluscs, suggested that modifications of that sort would have high selective value, and are to be expected. He gave one example, and Rothschild (1962) has given several more. We agree with Wright on the evolutionary probability of such modifications. There are a number of reports in the literature, not covered by Wright or Rothschild, which may be interpreted as examples of such modifications, and we are currently investigating other examples in our laboratory. Before examining these examples, however, it would be *apropos* to consider some current ideas about predator-prey relationships. In particular, we will examine some of the theoretical approaches which have identified important characteristics of predators, especially those characteristics which determine which prey individuals are taken. From this examination, we will try to deduce evolutionary strategies which parasites may adopt to increase the vulnerability of infected intermediate hosts to predation by the definitive host.

PREDATOR-PREY RELATIONSHIPS

Theoretical approaches

A substantial proportion of the ecologists studying predation are interested in the part predation plays in the population dynamics of the prey, therefore, in the responses predators display to changes in prey abundance. Responses to an increase in prey are generally an increase in the number of predators through immigration or reproduction (the numerical response) and an increase in the number of prey eaten per predator (the functional response). An analysis of the latter is the more instructive about the way predators act.

Holling (1961, 1965, 1966a, b) has developed progressively more complex mathematical models of the functional responses, with each step based on extensive experimental work. Holling focuses his attention on the *quantity* of prey eaten, not *which ones*. Therefore, he is not concerned with those factors of greatest interest to us. However, he has shown (1965) that as the density of one type of prey increases, vertebrate predators attack a higher proportion of that prey (to the limits of decreasing hunger), a pattern due to learning by the predator, or, as Tinbergen (1960) and others have put it, to the development of a search image. As will be shown later, search images appear to be important in the success of some of the behavioural changes induced by parasites.

Another approach to the study of predation is to deduce various aspects of an optimal evolutionary strategy of predators, draw inferences from that strategy, and determine whether or not the inferences are borne out by observations. Most of the authors using this approach have assumed that a basic aspect of an optimal strategy is for the predator to maximize the food value (energy content) obtained per unit time (or energy) expended in finding, catching and handling the prey (Emlen, 1966). Several observations (e.g.,

Tullock, 1971) suggest that predators do just that. The obvious inference for a parasite in an intermediate host (i.e., to minimize the time or energy required by the predator) is, of course, the topic of our paper.

Another, less obvious inference is also important: in the presence of a competitor, a predator should specialize the habitat used rather than its diet (MacArthur & Pianka, 1966; Schoener, 1969). This conclusion is substantiated by many observations on congeneric, competing species of birds (e.g., Haftorn, 1953, 1956a, b, c,; MacArthur, 1958; MacArthur & MacArthur, 1961). Even in waterfowl (the predators with which we work most extensively), which have been reported to have specialized diets (Olney, 1964; Bartonek & Hickey, 1969a, b; Dirschl, 1969; Bartonek & Murdy, 1970), much of the specialization appears to be due to different methods or locations of feeding (Collias & Collias, 1963; Sugden, 1969). Herting (1966) showed that the differences in foods of two species of predatory fishes were due to a combination of the behavioural patterns of the predators and those of the potential prey, and that "organisms highly susceptible to one predator may be nearly invulnerable to another" (p. 76). If habitat specialization, or specialization in another behavioural aspect of the feeding niche, is important, then it should be possible for parasites to modify the behaviour of their host in such a way as to increase the probability of being ingested, not only by predators in general, but specifically by their definitive hosts. Later in this paper we will give an example of such a modification.

A second aspect of an optimum strategy for a predator concerns *which individuals* should be taken. Slobodkin (1968) concluded that a predator normally should "manage" the prey so as to insure its continued availability, and that management consists of altering the natural pattern of (non-predatory) mortality as little as possible. The prudent predator should take those prey that are going to die anyway; in the words of the wildlife manager, predatory mortality should be compensatory. Is it? The literature on the subject is voluminous, and at times generates more heat than light. It is apparent that predation may be non-compensatory in some cases; it is also apparent that it frequently, perhaps usually, is compensatory (see reviews by Hirst, 1965; Errington, 1967). Slobodkin (1968) has pointed out that there should be strong selective pressure on the prey to make those animals most subject to predation the most expendable. Which are they, and can parasites take advantage of them for enhanced transmission?

Individual selection by predators

With a few exceptions, predators do not take different kinds of prey in direct proportion to their abundance; some kinds are taken in greater, others in lesser proportions. Craighead & Craighead (1956) used the term "vulnerability" to encompass factors which produce disproportionate usage. Others have used "preferred foods" to indicate those eaten in greater proportion than their abundance. However, it would appear desirable to use "preferred" only when based upon studies involving a free choice of alternate prey items, using "more vulnerable" as the less restrictive term.

One of the factors affecting vulnerability is the abundance of the prey in the precise feeding habitat used by the predator. Hornocker (1970) found that cougars, *Felis concolor* L., more frequently prey on wapiti, *Cervus canadensis* Erxleben, than on mule deer, *Odocoileus hemionus* (Rafinesque), despite the greater abundance of deer in the general area. Wapiti predominated in the bluff area in which the cougars were most successful in their hunting. Similar differences in the vulnerability of ungulates in different types of cover or terrain have been noted for the prey of tigers, *Leo tigris* (L.) (Schaller, 1967) and lions, *Leo leo* (L.) (Wright, B. S., 1960). The vulnerability of the amphipods *Hyalella azteca* Saussure to predation by mallards, *Anas platy-rhynchos* L., was very low because of the protection provided by the dense mats of vegetation the amphipods occupied (Perret, 1962). Similarly, Sugden (1965) found that the vegetation in his test aquaria protected the amphipod *Gammarus lacustris* Sars against predation by young lesser scaup, *Aythya affinis* Eyton. Bartonek & Murdy (1970) showed that conchostracans and chaoborids were highly vulnerable to younger scaup ducklings, but not to older ones, whereas amphipods and odonatan nymphs were vulnerable only to the older ducklings. They attributed the differences to a difference in the method of feeding. These observations suggest that parasites may be able to increase the vulnerability of their intermediate hosts by changing the latter's habitat selection.

On the basis of his work with muskrats, *Ondatra zibethicus* (L.), and quail, *Colinus virginianus* (L.), Errington (1967) hypothesized that, among territorial animals, predation is concentrated on those unable to establish territories. This "floating" or "excess" population generally occupies less favorable habitats, is often less well nourished, and often behaves differently from those occupying territories. This hypothesis has been supported by the work of Anderson (1961) on house mice, *Mus musculus* L., and Jenkins, Watson & Miller (1964) on red grouse, *Lagopus lagopus* L. Jenkins, *et al.* (1963) also found that the floating population of red grouse harbored more gastrointestinal parasites than those occupying territories. Thus far, however, no one seems to have investigated the possibility that the floating population might be instrumental in transmitting larval parasites to their predators.

Animals which differ in appearance are apparently more vulnerable to predation. There is an extensive literature (e.g., Ford, 1964; Carter, 1968) showing that animals that contrast with their backgrounds are more vulnerable. Pielowski (1959) showed that animals that differ from the rest of the population may also be more vulnerable: the hawks he observed preyed more extensively on pigeons that differed in colour than on the rest of the flock. Animals which differ in behaviour may also be more vulnerable. Wright, B. S. (1960 : 10) reported an instance of East African wild dogs, *Lycaon pictus* (Temminck), moving through a herd of Thomson's gazelles, *Gazella thomsonii* Gunther; the only one which appeared "panic-stricken at the sight of the approaching dog" was chased and killed. Estes & Goddard (1967) found that the leader of a pack of wild dogs picked out a victim, then ran it down. Their data suggest that some factors in the behaviour of the gazelle, possibly a reluctance to flee immediately, made certain individuals more vulnerable. Mech (1966 : 121-124) found that wolves, *Canis lupus* L., "tested" moose, *Alces alces* (L.), and rather quickly attacked or gave up. He assumed that the wolves could detect weakness by the behaviour of the moose. Similar testing of caribou, *Rangifer tarandus* (L.), by wolves was noted by Murie (1944), Banfield

(1954) and Crisler (1956). Fuller (1960) reported that bison, *Bison bison* (L.), generally showed no concern for wolves, even when very close. The only one that did show concern was a cow which had been wounded earlier. In the laboratory, both Mossman (1955) and Herting (1966) found that "disturbed" or "frightened" fish were more vulnerable to attack by a variety of predators. One of the factors that elicited a "fright response" was separation of the individual from the school (Herting, 1966). Mossman concluded that certain kinds of movements of the prey seem to act as sign stimuli for attack by visually-oriented predators.

Popham (1942) removed part of one or both hind legs of corixids, which then swam in circles or more slowly than normal and were more vulnerable to attack by rudd, *Scardinius erythrophthalmus* (L.). Herting & Witt (1967) found that impairment of the physical condition of various fish by the trauma of seining, a bacterial disease ("columnaris" disease), starvation, or parasitism by the monogenean, *Dactylogyrus* sp., resulted in sluggish movements and consequent greater vulnerability to predation by bowfins, *Amia calva* L.

Practically all workers who have investigated predation on birds or mammals have recognized the greater vulnerability of diseased or weak animals. Murie (1944), Crisler (1956), Fuller (1962) and Mech (1966) have shown disproportionate numbers of old, disabled or sick individuals in the prey of wolves; Borg (1962) noted that over half of the roe-deer, *Capreolus capreolus* (L.), taken by predators were injured or diseased; Rudebeck (1950, 1951) found that various predatory birds took injured or abnormal prey animals remarkably more frequently than such animals were seen in the field. Other examples have been reviewed by Hirst (1965) and Hornocker (1970). Rudebeck (1950, 1951) and others (reviewed by Hornocker, 1970) have pointed out that predators in these systems are relatively inefficient, capturing prey in well under half of their attempts. Mech (1970) summarized these results by saying, "The predator takes what it can catch."

One factor that decreases the efficiency of capture in some predator-prey systems is what Allee (1951) termed the "confusion effect", in which the predator is unable to single out a particular prey individual from a flock. An example of this effect, well-known among wildfowl hunters, is the tendency of a novice gunner to "flockshoot", and consequently hit nothing.

Schoener's (1969) studies suggest that inefficiency can be tolerated only when the prey is relatively large with respect to the size of the predator. On that basis, the greater efficiency of raptors feeding on insects (Rudebeck, 1951) or of ducklings feeding on various invertebrates (Collias & Collias, 1963; Sugden, 1969) is to be expected.

This difference in efficiency in different predator-prey systems would have important consequences for the strategies evolved by parasites to increase the probability of their transmission to the definitive host. Where the predator is an inefficient one, the strategy of the parasite could include decreasing the stamina of the prey, making it more conspicuous, or affecting its ability to respond to the predator. Where the predator is an efficient one, the strategy should be to make the prey more conspicuous.

Another factor having important consequences for the strategy of the parasite is the degree of overlap between the habitat occupied by the prey (the intermediate host) and the habitat (or feeding niche) of the predator (the definitive host). Where the habitat of the intermediate host is enclosed within the habitat of the definitive host (Fig. 1A), the strategic options of the parasite, as outlined in the preceding paragraph, are sufficient. However, when the habitats are only partly overlapping (Fig. 1B), the strategy is more complicated, and should also involve altering the behaviour of the infected intermediate host so that it occupies the area of overlap, as indicated by the arrow.

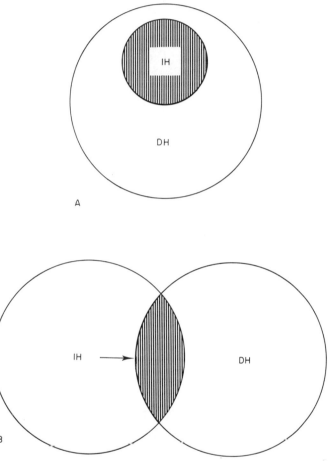

Figure 1. Where the habitat of the definitive host overlaps that of the intermediate host (A), varied strategies (see text) are open to the parasite. Where the habitats of definitive and intermediate hosts only overlap (B), the strategy should include altering the behaviour so that the infected intermediate host moves into the area of overlap.

In the balance of this paper, we will assess the evidence that parasites have in fact adopted those strategies.

Reduced stamina

The literature is replete with examples of helminths producing pathology in their intermediate hosts (e.g., reviews by Williams, 1967; Beck & Beverley-Burton, 1968; Smyth & Heath, 1970). However, a substantial portion of that literature deals with mortality or frank disease; very little attention has been paid to more subtle effects, particularly those on the behaviour of the infected intermediate hosts.

One system which has received some attention is that of the plerocercoid of *Ligula intestinalis* L. in the coelom of various fishes. Dence (1958) reported that infected shiners, *Notropis cornutus* Agassiz, were sluggish, less gregarious than uninfected fish, and frequented the shallower, warmer waters near shore, even when avian predators were near. When they did evade capture by retreating to deeper water, they soon headed back to shore, or assumed a course parallel to it. Orr (1966) found that infected rudd did not join spawning shoals, but were found in shallower waters, swimming near the surface. However, Arme & Owen (1968) found no evidence that the locomotor ability of infected roach, *Rutilus rutilus* (Flemming), was impaired, although under laboratory conditions, the infected fish died more rapidly than uninfected fish. In Alberta, we have noted infected spottail shiners, *Notropis hudsonius* (Clinton), and yellow perch, *Perca flavescens* (Mitchell), lagging behind schools of infected fish and often swimming closer to the surface. Various authors have suggested that these behavioural differences would make infected fish more vulnerable to predation by the definitive hosts, piscivorous birds, a suggestion substantiated by van Dobben (1952), who found that 30% of the roach in the diet of cormorants, *Phalacrocorax carbo* (L.) were infected, whereas only about 6.5% of the general population were infected.

The mechanisms underlying the behavioural changes have not been studied, although Whittaker & Feeny (1971) have suggested that an allomone (a chemical produced by one species to evoke, in another species, physiological or behavioural reactions which are favorable to the first species—Brown, Eisner & Whittaker, 1970) may be involved. Certainly, the physiological effects (reviewed by Williams, 1967; Arme & Owen, 1968), which include disturbances in the haematological picture and in the hormonal balance of the gonadal-hypophyseal axis, are suggestive of a biochemical disturbance.

Somewhat similar physiological disturbances are produced by *Schistocephalus solidus* (Muller) plerocercoids in the coelom of sticklebacks, *Gasterosteus aculeatus* L. (Williams, 1967; Arme & Owen, 1967), and infected fish are also sluggish and found near the surface, particularly on warm days (Clark, 1954; Arme & Owen, 1967; Lester, 1971). Baer (in Rothschild, 1962) reported that infected sticklebacks could easily be caught by hand. In Alberta, brook sticklebacks, *Culaea inconstans* (Kirtland), infected with the plerocercoids of *Schistocephalus* show the same behaviour and the same vulnerability to capture by hand, and presumably by piscivorous birds. Lester (1971) studied the distribution of infected sticklebacks in two lakes in British Columbia. In one, both infected and uninfected fish were found in shallow water during the breeding season (May to August), but only infected fish were found there during the rest of the year. In the other lake, 88% of a sample of fish taken near shore were infected, but only 12% of an offshore sample were infected. Lester suggested that the behaviour of the infected sticklebacks decreased their vulnerability to trout, which were never found in the shallow stream outlets occupied by large numbers of infected sticklebacks, but increased their vulnerability to piscivorous birds (the definitive hosts), which fed extensively in such areas.

Lester also found that heavily infected fish would not swim continuously at speeds over one body length per second, that infected fish used more oxygen than uninfected fish when at rest, and that the oxygen consumption of infected fish rose very rapidly with an increase in speed. He suggested that the behaviour of the infected fish may be due to their seeking well-oxygenated water and attempting to avoid excessive activity. His observations do not appear to agree with those of Walkey & Meakins (1970), who found that respiratory rates of infected sticklebacks were somewhat higher, but not significantly so, than those of uninfected fish, but that infected fish had markedly altered energy budgets and were more susceptible to starvation. However, the apparent disagreement is probably due to the way in which respiratory rates of infected fish were expressed. Lester expressed them in terms of fish tissue, whereas Walkey & Meakins expressed theirs in terms of fish plus parasite.

Another system which has attracted some attention, primarily of wildlife biologists, is that of *Echinococcus granulosus* (Batsch) in the lungs or liver of various ungulates, especially moose˙or caribou. Gross- and histo-pathology of the hydatids have been studied extensively (see reviews by Smyth, 1964, 1969; Smyth & Heath, 1970). Rausch (1952) examined a large number of infected moose in Alaska, and concluded that they showed little or no evidence of poor health. However, Fenstermacher (1937) regarded hydatids in moose as especially debilitating, Cowan (1951) considered that infected wapiti were usually impoverished and of low vitality, and Ritcey & Edwards (1958) described a moose which showed evidence of distress and collapsed when trapped for tagging; it harbored 30 golf-ball sized hydatid cysts. Summarizing these observations, plus his own on wolf-killed moose, Mech (1966) concluded that heavily infected animals are more vulnerable to wolves, the normal definitive hosts.

Various helminths inhabiting the musculature of their intermediate hosts have been considered debilitating to their hosts. Linton (1906) found that butterfish, *Poronotus (=Peprilus) triacanthus* (Peck) heavily infected with plerocercoids of *Otobothrium crenacolle* Linton, weighed less than lightly infected butterfish and concluded that the plerocercoids adversely affected the vitality of the fish. Grasshoppers, *Melanoplus* spp., infected with the larvae of the nematode *Tetrameres americana* Cram were "droopy and inactive, a condition that would make them easy prey for food-seeking fowls in nature" (Cram, 1931 : 4). Berland (1961) and Margolis (1970) reviewed literature on nematodes that infect the body musculature of fish; several produce extensive pathology to the muscle tissue, and some have been reported to make the host sluggish and vulnerable to predators. Locke, De Witt, Menjie & Kerwin (1964) gave indirect evidence for greater vulnerability to predation of forage fishes infected with the larvae of the nematode *Eustrongylides* sp. They found that heavy infections of *Eustrongylides* larvae migrating through the viscera had killed 36 red-breasted mergansers, *Mergus serrator* L., which had been observed feeding on dead and weakened "minnows". They also found 6 of 40 silversides, *Menida beryllina* (Cope), and 12 of 219 mosquitofish, *Gambusia affinis* (Baird & Girard), collected from the same pond, infected with the nematode larvae. The disparity between the large numbers of nematodes found in the mergansers and those found in the fish suggests that the infected fish were more vulnerable to the mergansers. Leiby & Dyer (1971 : 202) indicated that heavy infections of *Taenia multiceps* Leske in lagomorphs hinder movement, increasing their vulnerability to the canid definitive hosts. Sindermann (1970 : 314) concluded that, "Examinations of fish ... with heavy larval nematode infestation of muscles ... lead inevitably to the conclusion that such a parasite burden, although it may not be the primary cause of death, must seriously reduce statistical chances for survival of the host in an environment in which early and sudden death is the rule." His argument appears to hold for the other host-parasite systems as well.

Sogandares-Bernal & Lumsden (1964) described heavy infections of metacercariae of *Ascocotyle leighi* Burton, which are found under the luminal endothelium of the bulbus arteriosus of some poeciliid and cyprinodont fishes, in which the bulbus was almost completely blocked by the cysts. Although they found that the general appearance and behaviour of heavily infected fish could not be distinguished from that of uninfected fish, heavy accidental mortality in one laboratory tank was limited to fish with "upwards of 80 cysts per bulbus, whereas the ones remaining alive were infected with significantly fewer metacercariae, usually two or three cysts" (pp. 10—11). Hopkins (in Sindermann, 1970) collected, with a dip net, a sample of sheepshead minnows, *Cyprinodon variegatus* Lacepede, that was 100% infected with metacercariae of an *Ascocotyle (leighi?)* in the much enlarged bulbus. A sample collected later in the same area with a seine was only about 10% infected, suggesting that the infected fish were more susceptible to dip-net predation, and presumably that by the avian definitive hosts.

Laboratory tests of the stamina of infected intermediate hosts are almost non-existent. Except for the study of Lester (see above), the few which have been done have not involved systems for which there are field observations suggestive of significantly increased vulnerability to predation. Three laboratory studies (von Brand, Weinstein & Wright, 1954; Bernard, 1959; Goodchild & Frankenberg, 1962) have shown reduced stamina and reduced voluntary activity in rodents infected with *Trichinella spiralis* (Owen). Fox (1965) showed a reduced tolerance to high water temperatures, reduced haematocrit values and decreased stamina in swimming tests in rainbow trout, *Salmo gairdneri* Richardson, infected with metacercariae of *Bolbophorus confusus* (Krause) in the musculature. Butler (in Milleman & Knapp, 1970) stated that coho, *Oncorhynchus kisutch* (Walbaum), and rainbow trout show retarded growth and impaired swimming performance when infected with the metacercariae of *Nanophyetus salmincola* (Chapin). Olson (1968), however, could find no differences in swimming performance, haematocrit value or tolerance to high temperatures or low oxygen levels between rainbow trout infected with the metacercariae of *Cotylurus erraticus* (Rudolphi) and uninfected trout. (*C. erraticus* metacercariae encyst in the pericardial cavity.)

Only one investigator (Coble, 1970) has tested, in the laboratory, the vulnerability of an infected intermediate host to predation. Fathead minnows, *Pimephales promelas* Rafinesque, infected with metacercariae of *Clinostomum marginatum* (Rudolphi), and controls, were placed in a tank with large-mouth bass, *Micropterus salmoides* (Lacepede). He found no differences in the proportion of the infected and uninfected minnows eaten by the bass. It should be noted, however, that most *Clinostomum* metacercariae are subcutaneous and there is little evidence that they adversely affect their host (Klaas, 1963). In addition, bass (which are not definitive hosts) feed in an entirely different manner from herons (which are). Coble's experiment is therefore not an adequate test of our hypothesis.

Increased conspicuousness

Most of the examples of increased conspicuousness produced by parasites in their intermediate hosts are not attributable to the behaviour of the infected host, but to some morphological attribute of it. However, increased conspicuousness is frequently associated with other changes in the behaviour of the host, and may help to attract visually-oriented predators. "Black spot", produced by accumulations of pigment around the metacercariae of certain heterophyid or strigeid trematodes, was regarded by Rothschild (1962) as increasing the vulnerability of the infected fish to predation by making it more conspicuous to the avian predator. Presumably, the same would apply to the large yellow cysts of *Clinostomum marginatum.*

The swollen abdomens of fish infected by plerocercoids of *Ligula* or *Schistocephalus* show as white stripes in a dorsal, "predator's eye" view. Mice infected with the spargana of *Spirometra* spp. become obese (Mueller, 1966). Lagomorphs heavily infected with the coenuri of *Taenia multiceps* may be retarded in (or prevented from) changing fur color (Leiby & Dyer, 1971). The cystacanths of various species of polymorphid acanthocephalans are orange and easily seen through the body wall of the amphipod host (Denny, 1969; Podesta & Holmes, 1970). The marine copepod *Calanus finmarchicus* (Gunnerus) is reddish and quite distinctive when infected with metacercariae of *Derogenes varicus* (Muller) (M. Weinstein, pers. comm.). All of these features make the infected animal more conspicuous and perhaps more vulnerable to predation.

More interesting are the modifications in terrestrial or amphibious snails infected with the highly modified sporocysts, the broodsacs, of *Leucochloridium* or *Neoleucochloridium* (reviewed by Ulmer, 1971: 136-137). The broodsacs, which have transverse bars of red, brown or green, invade the tentacles of the snail, deform them, and pulsate in response to light. This behaviour of the parasite is combined with an altered response of the host (described below) to make the parasitized snail very conspicuous. Thus far, however, no one has actually tested the effect of this conspicuousness on predation by the avian hosts.

Disorientation

Disorientation of infected intermediate hosts appears to be due to extensive pathology in the central nervous system or the major sensory receptors of the host. Disorientation may have three interacting effects on the behaviour of the host — the host may be more conspicuous, it may wander into unusual habitats, and its escape responses to predators may be disturbed or obliterated.

Strigeatoid metacercariae localizing in the lens of the eyes of various fishes are known to produce opacity of the lens (parasitic cataract) and blindness (reviewed by Dubois, 1944; Dogiel, Petrushevsky & Polyanski, 1961; Larson, 1965; Williams, 1967). Pathology appears to be limited to heavy infections and is most frequent in hatcheries or rearing ponds. In observations on forage fishes in Alberta, we have never encountered a heavily infected fish, nor any signs of damage to the eyes.

Metacercariae in the brain of fishes are more pathogenic. Szidat & Nani

(1951) found that metacercariae of *Austrodiplostomum mordax* Szidat & Nani develop in the ventricles of pejerrey, *Basilichthys* spp., and attack the surrounding tissue only when they reach an infective stage. The cerebellar tissues and optic centers of infected fish are severely damaged; the blind, incoordinated fish are tumbled about on the surface of the water, completely inactive and showing no evasive behaviour (Szidat, 1969). Evidence of similar systems have been given by Hoffman & Hoyme (1958) for *Diplostomum baeri eucaliae* Hoffman & Hundley in brook sticklebacks, by Lautenschlager (1959) for *Diplostomulum* sp. in newts, *Triturus (= Diemictylus) viridescens* (Rafinesque), and by Szidat & Nani (1951) for *Tylodelphys destructor* Szidat & Nani in pejerrey and other fishes. We have found metacercariae of *Tylodelphys podicipina* Kozicka & Niewiadomska in the ventricles of spottail shiners in Alberta; some, but not all of the infected fish showed the disoriented behaviour described by Szidat (1969) and similar, although slightly different, neural pathology.

The coenuri of the tapeworm *Taenia multiceps* develop in the brain and spinal cord of sheep and other ruminants, producing a disease known as "Gid", characterized by symptoms such as circling, staggering, and loss of appetite; infected sheep often fall behind the herd and become lost. Becklund (1970) has concluded that this parasite has probably become extinct in the United States after years of predator control practices and modern sheep management. It is interesting to note that wolves open the brain case of prey up to the size of a calf moose or caribou when feeding on their kills (Burkholder, 1959).

Several examples of nematode larvae in the central nervous systems of mammals, including cases causing disturbed orientation, are given by Sprent (1955a, b; 1962) and Anderson (1968). The best example is the study by Tiner (1953a, b; 1954) on the larvae of *Ascaris columnaris* Leidy in mice and squirrels. Larvae migrating in the central nervous system produced incoordination, blindness, loss of fear to larger animals, and eventual death in both natural and experimental infections. Tiner stressed the significance of these effects to the life cycle of the parasite and related them to the feeding habits of the natural definitive hosts.

Altered responses

These are the most intriguing behavioural modifications, and the ones most conducive to laboratory investigation. They are not due to pathology to the nervous system, but to changed responses to certain environmental stimuli.

The best examples of such effects involve ants and the metacercariae of dicrocoelid trematodes. The effects of *Dicrocoelium dendriticum* have already been discussed. Carney (1969) discovered several behavioural peculiarities in carpenter ants, *Camponotus* spp., carrying metacercariae of *Brachylecithum mosquensis* (Skriabin & Isaichikov). Infected ants contained at least one metacercaria in or around the supraoesophageal ganglion and several others in the tissues of the markedly enlarged gaster. A much higher proportion (91%) of infected worker ants were found in a collection made in an open, rocky area than in a collection from an adjacent wooded area (9%). Infected ants could typically be found slowly circling, or remaining motionless for hours on the surfaces of rocks. Infected ants did not respond to sudden changes in light intensity produced by shading them. Carpenter ant workers in temperate regions are normally strongly photophobic (Wheeler, 1910). Tapping the rock on which an infected ant was circling or resting produced a "brief stirring. The ant would return to its previous pattern of behaviour within a short time, however." (Carney, 1969: 608). The response of infected ants to temperature may also be altered, since Carney found only infected ants active outside of the nests in late fall. These behavioural modifications would obviously increase the vulnerability of the infected ants to predation by insectivorous birds, including robins, the definitive hosts for *B. mosquensis.*

Another metacercaria, a plagiorchioid, was found by Lewis & Wright (1962) near the brain of a high proportion of blackflies, *Simulium exiguum,* collected as pupae, but not in any of 585 collected while biting humans in the same area. Rothschild (1962) suggested that their absence from the biting population may be due to a sluggish and unwary behaviour in infected flies, leaving them more vulnerable to predation.

Graham (1966) briefly described behavioural changes in the beetle, *Tribolium confusum* Duv., when infected with the cysticercoid of the chicken tapeworm, *Raillietina cesticillus* (Molin). Infected beetles do not show the highly photophobic tendencies displayed by normal beetles and do not seek concealment, hence increasing the risk of predation by chickens.

Swennen (1969) has reported unusual behaviour in bivalve molluscs, *Macoma balthica* (L.), infected with sporocysts containing metacercariae of a gymnophallid trematode. *Macoma* normally remain well concealed, buried 1 to 12 cm into the sediment; their retracted siphons usually leave no hole. "It is certainly not a prey which can be easily traced by a predator" (Swennen, 1969: 378). The infected *Macoma,* however, crawl along, just under the surface of the sandy tidal flats in the high intertidal zone, leaving conspicuous tracks which indicate their position (Plate 1A). In a survey, Swennen found 100% infection in two samples of *Macoma* recovered at the ends of the crawling tracks, as compared to 13% and 5% infection in clams from sieved random bottom samples from the same locations. Swennen described the crawling behaviour as a waste of energy and an increased risk of predation by visually-hunting shorebirds, gulls or waterfowl, normal definitive hosts of gymnophallids (Yamaguti, 1958).

According to Sparks & Chew (1966), littleneck clams, *Venerupis* (= *Protothaca*) *staminea* (Conrad), heavily infected with plerocercoids of the tetraphyllidean cestode *Echeneibothrium* sp. were found exposed on a gravel bed, rather than in the typical buried position. This would render them more vulnerable to predation by molluscivorous skates or rays, the definitive hosts. There were often more than 35 plerocercoids in a single section through an infected clam. Warner & Katkansky (1969) found a maximum intensity of eight *Echeneibothrium* spp. plerocercoids in the littleneck clams they were studying; infected clams were buried in the mud and not exposed. The authors, however, did not disagree with Sparks & Chew (1966) that heavy infections such as the latter found could produce abnormal exposure on the surface.

The increased conspicuousness of snails infected with broodsacs of *Leucochloridium* and *Neoleucochloridium* has been mentioned above. In addition, however, Wesenberg-Lund (1931) has pointed out that infected snails seem to seek light, unlike the uninfected ones, and are characteristically found in the open on marshy vegetation (Ulmer, 1971).

Schütze (in Lehmann, 1967) reported that *Gammarus pulex fossorum* Koch infected with the cystacanths of the acanthocephalan *Echinorhynchus truttae* Schrank occurred more frequently in drift than would be expected from their prevalence in the overall population. Crompton (1970) attributes this to a weakening of the infected gammarids, but Lehmann (1967: 259) clearly states that it must not be regarded as a sign of weakening of the gammarids,

> "Dass Gammariden, die mit Larven von *Echinorhynchus truttae* befallen sind, haufiger in der Drift auftreten, als ihrem Anteil an der Population entspricht (Abb. 18/4), darf nicht als Anzeichen einer Schwächung durch den Parasiten aufgefasst werden."

Based on our unpublished investigations on gammarids infected with other acanthocephalans, it would appear that an altered response to light, and perhaps current, might be involved. The work of Jenkins, Feldmeth & Elliott (1970) suggests that the increased prevalence in the drift would make the infected gammarids more vulnerable to predation by trout (*Salmo* spp.), the definitive hosts.

The observations summarized in this section strongly suggest relationships between altered responses of the infected intermediate hosts and greater vulnerability to predation by the definitive hosts. However, in no case has either the specific response of the infected intermediate host or the actual vulnerability to predation been subjected to critical experimental evaluation.

BEHAVIOUR OF *GAMMARUS LACUSTRIS* INFECTED WITH *POLYMORPHUS PARADOXUS*

Polymorphus paradoxus (Connell & Corner) was first described from beavers and muskrats collected in Alberta. Workers in our laboratory have also found ovigerous adults in mallard ducks. *Gammarus lacustris* Sars is the intermediate host (Denny, 1969).

Denny (1967) and others working at Cooking Lake, Alberta, a major study area for our group, noticed that gammarids infected with *P. paradoxus* were often attached to or closely associated with floating material. Preliminary tests and observations showed that infected gammarids were actually clinging to material on or near the surface of water, and that they remained clinging even after being disturbed. These behaviour patterns were not seen in uninfected gammarids.

The system appeared to be an excellent one in which to test the hypothesis that altered responses of the infected gammarids to environmental stimuli greatly increase the vulnerability of the infected gammarids to selective predation by the definitive hosts. In this section, we report the results of experiments designed to test whether such altered responses exist and whether they increase the vulnerability of infected gammarids to predation by mallard ducks.

Materials and methods

Collection and identification of material

Infected gammarids were collected from random samples taken at Cooking and Hastings Lakes, Alberta, or from floating objects in areas where other work was being carried out. No difference was found in the behaviour of infected gammarids collected by the two methods. Cystacanths of *P. paradoxus* are easily detectable through the carapace of living gammarids (Plate 1B), and can be easily distinguished from other local polymorphid cystacanths by their large size.

Description of tests

Different groups of gammarids composed of equal numbers of infected and uninfected animals were used for each test.

Light stimulus tests. One infected and one uninfected gammarid were placed in an aquarium divided into equal light and dark areas. The only light in the environmental control room was from a lamp placed directly over the aquarium. An equal amount of floatage was provided in each zone.

The gammarids to be tested were introduced into the observation aquarium in total darkness and allowed at least five minutes to settle; the light was then switched on and their responses were recorded on a keyboard operated, multiple channel event recorder. The frequency and duration of time spent in the dark zone, in the lower part of the lighted zone, in the upper 3 cm of the lighted zone but not clinging, and clinging to the floatage were recorded for each gammarid. Each observation lasted for one hour. In order to distinguish between the effects of light and gravity, one series of experiments was run in the dark. The aquarium containing infected and uninfected gammarids was disturbed, then allowed to settle. The light was switched on and the position of the gammarids recorded. In a second series, light sources were positioned at one side or through one half of the bottom of an aquarium. In the latter test, the other half of the aquarium was darkened and provided with a mud bottom.

Tests on clinging behaviour. Infected and uninfected gammarids (up to 175 each) were exposed to floating wood, reeds, and *Potamogeton* sp. in an aquarium in the laboratory at room temperature, 18°-20° C, with normal overhead lighting. The numbers of gammarids clinging to the three types of floatage and to the corners of the aquarium (in the upper 3 cm of water only) were recorded three times daily. Floatage had to be picked out of the water in order to distinguish whether gammarids were showing the clinging behaviour or just resting temporarily. This disturbance in the water was enough to test the behaviour of gammarids in the corners, since uninfected gammarids will release and dive to the bottom; these reactions were confirmed by earlier tests and observations. The countings were spaced at least two hours apart to allow gammarids to resettle.

Predator tests. One or two mallard ducks were used as predators and equal numbers (up to 75 each) of infected and uninfected gammarids as prey. The gammarids were allowed 10-15 minutes to settle after being introduced into the tank. The mallards were fed *ad libitum* on a mash diet, and were never starved before a test. Different mallards were used for each test.

Tests 1 to 4 were performed outside, in a tank measuring 5 foot 10 inches × 1 foot 11 inches × 1 foot deep; tests 5 and 6 were done in a laboratory observation room fitted with one-way glass, in a tank 4 foot × 2 foot 5 inches × 1 foot 10 inches deep. The tanks were lined with a fine mesh net to recover the surviving gammarids at the end of each test. A mud bottom and an equal amount of floatage were provided for each test. The temperature, turbidity, and pH were kept as consistent with the conditions in the study lakes as possible.

Results and discussion

Behaviour of gammarids

Uninfected gammarids were in the dark zone almost all the time; they were found in the lighted zone for only seven minutes out of ten hours of observation. They showed no preference for the upper part of the lighted zone (Fig. 2).

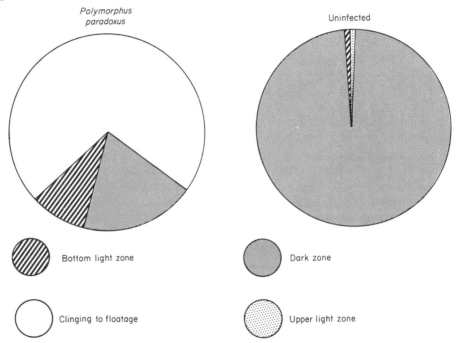

Figure 2. Proportion of time spent in various habitats by *Gammarus lacustris* infected with *Polymorphus paradoxus* and by control gammarids.

Infected gammarids, however, spent over 80% of their time in the lighted zone. They showed a distinct preference for the upper part of that zone, spending over 70% of the total time there; all of this time was spent clinging to the floatage (Fig. 2).

These results can be expressions of a positive phototaxis or a negative geotaxis. We distinguished between them with two series of experiments. In the dark, both infected and uninfected gammarids were found on the bottom. When the light was directed from the side or from below the aquarium, uninfected gammarids immediately swam away from the light, oriented with their dorsal side toward the light source; when the light was from below the aquarium, they swam diagonally upwards, toward the dark zone. On reaching the dark zone, they immediately turned over so that their dorsal side was uppermost, and swam towards the bottom. The infected gammarids, however, immediately swam towards the light, orienting with their dorsal side toward the light source, and congregated in the area of most intense light. These results indicate that the basic response is a positive phototaxis.

In a separate series of experiments, designed to test the preference of infected and uninfected gammarids for different substrates, the results of the previous experiments were reinforced. The only gammarids showing the clinging behaviour were those infected with *P. paradoxus*; 60% of these were clinging when counted. They seemed to prefer clinging onto the rough ends of reeds (Fig. 3), sometimes numbering up to 40 on one piece of reed. Fewer gammarids were clinging to the corners of the tanks or to other types of floatage.

Infected gammarids "cling" with their gnathopods, and usually remain locked onto material even after the latter is shaken (Plate 2). They often have to be forcibly pulled off, sometimes losing a gnathopod in the process. When they are dislodged, they skim along the surface of the water and cling onto the first suitable object. "Skimming" is so pronounced that it almost appears to be

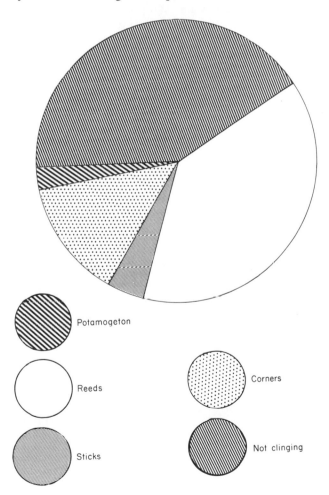

Figure 3. Proportions of *Gammaris lacustris* infected with *Polymorphus paradoxus* clinging to various substrates.

an effort to get out of the water and causes an obvious surface disturbance. The gammarids appeared to be caught in the surface tension, but when pushed under the water surface, they immediately swam back up to the surface and resumed skimming.

In the laboratory, all of these modified behavioural patterns of infected gammarids are manifested following a disturbance. Instead of diving to the bottom, as uninfected gammarids do at any disturbance, infected gammarids are attracted by their positive phototaxis to the surface, where the other two behavioural abnormalities, skimming and clinging, take place. The entire reaction appears to be an abnormal evasive response.

In the field, the behaviour patterns tend to segregate the infected from the uninfected gammarids. The extensity of *P. paradoxus* in randomly collected samples of gammarids from our main study areas at Cooking and Hastings Lakes is very low. For example, the extensities in Cooking Lake in July and August, 1969 were 4.1% and 0.9% (J. Tokeson, pers. comm.), yet up to 35 infected gammarids could be collected from each of several dead waterfowl which had been floating in the same sampling sites for only 10-20 minutes. Even greater differences between population extensities and numbers of clinging gammarids were noted at Hastings Lake in 1970. Hundreds of infected gammarids could be found clinging along the waterline of a wooden boat dock or on floating debris when random population samples taken at the same time (but in a different bay) showed *P. paradoxus* extensities of 0% (July) and 0.4% (September).

Predator tests

In all tests mallards were attracted to the floating material and fed around the floatage before dabbling or tipping. Since all nine mallards behaved in the same way, and all of them were laboratory-raised and fed completely on a mash diet, this method must be one of their innate feeding mechanisms (Weidmann, 1956). When the floating material was disturbed, most of the gammarids remained clinging, although a few skimmed away along the surface. Mallards fed first on the clinging gammarids, then on any skimming in the immediate vicinity. The mallards returned to the floatage repeatedly, turned it over and fed on any gammarids present. They also struck at any small objects (including bubbles) on the water surface. These last reactions appear to be due to the formation of a search image.

Mallards are dabbling ducks, and feed chiefly by dabbling on the water surface. Mallards pick out initial food items by visual means (Perret, 1962; our observations) and apparently feed underwater only when surface items are not available. Tipping may also be an important feeding method, especially in shallow areas where they are able to reach the bottom. According to Weidmann (1956), the various feeding movements are activated by the degree of hunger. Our observations during the predator tests imply that the order of appearance of their different mechanisms depend a great deal on the accessibility of food, and on search images which may have been formed earlier.

On two occasions, infected gammarids were seen clinging onto the feathers of ducks following the predator tests. It is possible that mallards ingest considerable numbers of infected gammarids while preening.

The combination of the different behaviour patterns of infected and uninfected gammarids, and the feeding behaviour of mallards, resulted in a disproportionately large number of infected gammarids being eaten in each of the six predator tests. Over four times as many infected gammarids were eaten (Table 1). Chi-square contingency tables showed that the differences in the proportions eaten between infected and uninfected gammarids were highly significant. Gammarids infected by cystacanths of *Polymorphus paradoxus* are more vulnerable to predation by mallards, one of the definitive hosts of this acanthocephalan.

Table 1. Vulnerability of gammarids infected with *Polymorphus paradoxus* to predation by mallard ducks

Test No.	Duration (min)	No. ducks	Gammarids eaten		P
			Uninfected	Infected	
1	7	2	6/25*	16/25	<0.005
2	5	2	13/50	35/50	<0.0005
3	5	2	12/50	42/50	<0.0005
4	5	1	8/50	18/50	<0.025
5	10	1	0/75	48/75	<0.0005
6	15	1	24/75	63/75	<0.0005
Total	47	9	53/325	222/325	<0.0005

* Number of gammarids eaten/number available

This vulnerability may be rather specific to mallards, since the behavioural studies of Collias & Collias (1963) and Sugden (1969) indicate that various species of waterfowl use different methods of feeding. We are currently investigating this specificity. The behavioural modifications of the infected gammarids would also appear to make them more vulnerable to muskrats and beaver, the other definitive hosts of *P. paradoxus*. We are currently investigating this possibility as well.

DISCUSSION

It should be clear from this review that, on this topic, the number of speculations greatly exceeds the number of investigations. Anecdotal reports of behavioural effects are useful, and we have had to rely heavily on them. However, anecdotal reports suggesting no behavioural effects are very difficult to interpret, since behavioural effects are sometimes subtle and existing data are insufficient to determine what degree of effect is necessary to produce added predation.

The data are insufficient to assess the extent to which the strategies outlined above (see Predator-prey relations) are used by parasites, but that they are used is obvious. The observations on the effects of *Ligula, Schistocephalus,* and others, strongly suggest that they decrease the stamina of their intermediate hosts with consequent behavioural changes and increased vulnerability to predation by the definitive hosts. *Leucochloridium,* and probably other parasites as well, do make their hosts more conspicuous. Szidat's strigeids and Tiner's ascarids definitely interefere with their intermediate hosts' responses to predators. Finally, dicrocoelid metacercariae in ants, gymnophallid metacercariae in clams, and cystacanths of *Polymorphus paradoxus* in gammarids, alter the behavioural responses of their hosts in such a way as to move infected intermediate hosts into the zone of overlap with the feeding niche of the definitive hosts.

Of course, not all parasites use the same strategies, and there are many parasites which apparently do not affect their intermediate hosts' behaviour (Sillman, 1957; Olson, 1968; Spall & Summerfelt, 1970).

Selective predation on infected individuals is of obvious advantage to the parasite, but is it not also advantageous to the prey population? Many of the best examples of effects on the behaviour of infected intermediate hosts involve parasites that do considerable damage to the host, or sterilize it, or both. The destruction of host gonads by plerocercoids of *Ligula* and *Schistocephalus* are well known; *Leucochloridium,* like many other trematodes, effectively castrates its host (Wesenberg-Lund, 1931); the extensive development of gymnophallid sporocysts in clams suggests that they, too, are castrated; our observations indicate that cystacanths of *Polymorphus paradoxus* also castrate their hosts. Szidat's strigeids and Tiner's ascarids inevitably kill their hosts. Behavioural changes produced by dicrocoelid metacercariae in ants obviously eliminate those individuals as productive members of the ant society. Therefore, predation on infected individuals is predation on the excess, expendable individuals. This reverses the process suggested by Slobodkin (1968), and makes the expendable individuals the ones most vulnerable to predation.

In addition, there appears to be another advantage accruing to the prey population. If there are sufficient infected individuals, predators might be expected to develop search images dependent upon the behavioural peculiarities of those infected individuals (as in the mallards in our predator tests). Such search images would be protective of the reproductives in the prey population.

Szidat (1969 : 763) gave the following quotation from an unpublished manuscript (1933) of the late George R. LaRue:

> "In gaining an entrance into their hosts the parasitic worms seem to show the most astounding knowledge of the activities and habits of life of the host. Had they ability to see, hear, and reason it seems doubtful whether they could exhibit a more diabolical cunning to gain their ends than they do now."

Much of this "diabolical cunning", due, of course, to long evolutionary adaptation, shows to best advantage in the behaviour of parasites, or the behaviour they induce in their hosts. As evidenced by this symposium, the study of parasite behaviour, which Szidat (1969 : 784) feared was non-existent, is attracting increasing attention. We can find no better way of concluding this paper than to quote Martin Ulmer (1971 : 147):

> "The study of behaviour has not yet developed as a recognized area in helminthology, yet accumulating evidence suggests more and more that in every early stage, highly specialized, complex behaviour patterns occur in

response to exacting requirements of each species. The careful and critical analysis of adaptive behaviour for each life cycle stage, and the elucidation of trigger mechanisms including chemical, hormonal, sensory, and neurosensory stimuli, undoubtedly will provide challenging areas of inquiry for the intellectually curious helminthologist."

ACKNOWLEDGEMENTS

The work on *Polymorphus paradoxus* will form part of a dissertation by the junior author. We would like to thank other members of the parasitology group at the University of Alberta for helpful discussions and unpublished data. J. L. Mahrt, T. G. Neraasen and W. M. Samuel criticized the manuscript. This investigation was supported in part by National Research Council of Canada Operating Grant A-1464 to the senior author.

REFERENCES

ALLEE, W. C., 1951. *Cooperation among animals.* New York: Schuman.

ANDERSON, P. K., 1961. Density, social structure, and nonsocial environment in house-mouse populations and the implications for regulation of numbers. *Trans. N.Y. Acad. Sci., 23:* 447-451.

ANDERSON, R. C., 1968. The pathogenesis and transmission of neurotropic and accidental nematode parasites of the central nervous system of mammals and birds. *Helminth. Abstr., 37:* 191-203.

ANOKHIN, I. A., 1966. Daily rhythm in ants infected with metacercariae of *Dicrocoelium lanceatum. Dokl. Akad. Nauk SSSR, 166:* 757-759.

ARME, C. & OWEN, R. W., 1967. Infections of the three-spined stickleback, *Gasterosteus aculeatus* L., with the plerocercoid larvae of *Schistocephalus solidus* (Muller, 1776), with special reference to pathological effects. *Parasitology, 57:* 301-314.

ARME, C. & OWEN, R. W., 1968. Occurrence and pathology of *Ligula intestinalis.* Infections in British fishes. *J. Parasit., 54:* 272-280.

BANFIELD, A. W. F., 1954. Preliminary investigations of the barren-ground caribou. II. Life history, ecology and utilization. *Wildl. Mgmt. Bull., Ottawa (Ser.), 1(10b).*

BARTONEK, J. C. & HICKEY, J. J., 1969a. Food habits of canvasbacks, redheads, and lesser scaup in Manitoba. *Condor, 71:* 280-290.

BARTONEK, J. C. & HICKEY, J. J., 1969b. Selective feeding by juvenile diving ducks in summer. *Auk, 86:* 443-457.

BARTONEK, J. C. & MURDY, H. W., 1970. Summer foods of lesser scaup in subarctic taiga. *Arctic, 23:* 35-44.

BECK, J. W. & BEVERLEY-BURTON, M., 1968. The pathology of *Trichuris, Capillaria* and *Trichinella* infections. *Helminth. Abstr., 37:* 1-26.

BECKLUND, W. W., 1970. Current knowledge of the gid bladder worm *Coenurus cerebralis* (= *Taenia multiceps*) in North American domestic sheep, *Ovis aries. Proc. helminth. Soc. Wash., 37:* 200-203.

BERLAND, B., 1961. Nematodes from some Norwegian marine fishes. *Sarsia, 2:* 1-50.

BERNARD, G., 1959. Experimental trichinosis in the golden hamster. I. Spontaneous muscular activity patterns. *Am. Midl. Nat., 62:* 396-401.

BORG, K., 1962. Predation on roe-deer in Sweden. *J. Wildl. Mgmt., 26:* 133-136.

BRAND, T. VON, WEINSTEIN, P. P. & WRIGHT, W. H., 1954. The working ability of rats infected with *Trichinella spiralis. Am. J. Hyg., 59:* 26-31.

BROWN, W. L., EISNER, T., JR. & WHITTAKER, R. H., 1970. Allomones and kairomones: Transspecific chemical messengers. *Bioscience, 20:* 21-22.

BURKHOLDER, B. L., 1959. Movements and behaviour of a wolf pack in Alaska. *J. Wildl. Mgmt., 23:* 1-11.

CARNEY, W. P., 1967. *The life history of* Brachylecithum mosquensis *(Skriabin & Isaichikov, 1927) (Trematoda : Dicrocoeliidae).* Ph.D. thesis, University of Montana, Missoula.

CARNEY, W. P., 1969. Behavioral and morphological changes in carpenter ants harboring dicrocoeliid metacercaria. *Am. Midl. Nat., 82:* 605-611.

CARTER, M. A., 1968. Thrush predation of an experimental population of the snail *Cepaea nemoralis* (L.). *Proc. Linn. Soc. Lond., 179:* 241-249.

CLARK, A. S., 1954. Studies on the life cycle of the pseudophyllidian cestode *Schistocephalus solidus. Proc. zool. Soc. Lond., 124:* 257-302.

COBLE, D. W., 1970. Vulnerability of fathead minnows infected with yellow grub to largemouth bass predation. *J. Parasit. 56:* 395-396.

COLLIAS, N. E. & COLLIAS, E. C., 1963. Selective feeding by wild ducklings of different species. *Wilson Bull., 75:* 6-14.

COWAN, I. McT., 1951. The diseases and parasites of big game mammals of western Canada. *Proc. Fifth Ann. Game Conv.,* pp. 37-64.

CRAIGHEAD, J. J. & CRAIGHEAD, F. C., JR., 1956. *Hawks, owls, and wildlife.* Harrisburg, Pa.: Stackpole Co.

CRAM, E. B., 1931. Developmental stages of some nematodes of the Spiruroidea parasitic in poultry and game birds. *Technical Bulletin, U.S. Department of Agriculture, Washington, D.C., 227:* 1-27.

CRISLER, L., 1956. Observations of wolves hunting caribou. *J. Mammal., 37:* 337-346.

CROMPTON, D. W. T., 1970. *An ecological approach to acanthocephalan physiology. Cambridge Monographs in Experimental Biology.* Cambridge: University Press.

DENCE, W. A., 1958. Studies on *Ligula*-infected common shiners (*Notropis cornutus* Agassiz) in the Adirondacks. *J. Parasit., 44:* 334-338.

DENNY, M., 1967. *Taxonomy and seasonal dynamics of helminths in* Gammarus lacustris *in Cooking Lake, Alberta.* Ph.D. thesis, University of Alberta, Edmonton.

DENNY, M., 1969. Life cycles of helminth parasites using *Gammarus lacustris* as an intermediate host in a Canadian lake. *Parasitology, 59:* 795-827.

DIRSCHL, H. J., 1969. Food of lesser scaup and blue-winged teal in the Saskatchewan River Delta. *J. Wildl. Mgmt., 33:* 77-87.

DOBBEN, W. H. VAN, 1952. The food of the cormorant in the Netherlands. *Ardea, 40:* 1-63.

DOGIEL, V. A., PETRUSHEVSKI, G. K. & POLYANSKI, YU. I., 1961. *Parasitology of fishes.* Edinburgh & London: Oliver & Boyd.

DUBOIS, G., 1944. A propos de la spécificité parasitaire des Strigeida. *Bull. Soc. neuchâtel. Sci, nat., 69:* 5-103.

EMLEN, J. M., 1966. The role of time and energy in food preference. *Am. Nat., 100:* 611-617.

ERRINGTON, P. L., 1967. *Of predation and life.* Ames: Iowa State University Press.

ESTES, R. D. & GODDARD, J., 1967. Prey selection and hunting behaviour of the African wild dog. *J. Wildl. Mgmt., 31:* 52-70.

ETGES, F. J., 1963. Effects of *Schistosoma mansoni* infection on chemosensitivity and orientation of *Australorbis glabratus. Am. J. trop. Med. Hyg., 12:* 696-700.

FENSTERMACHER, R., 1937. Further studies of diseases affecting moose. II. *Cornell Vet., 27:* 25-37.

FORD, E. B., 1964. *Ecological genetics.* New York: Wiley.

FOX, A. C., 1965. *The life cycle of* Bolbophorus confusus *(Krause, 1914) Dubois, 1935 (Trematoda : Strigeoidea) and the effects of the metacercariae on fish hosts.* Ph.D. thesis, Montana State University, Bozeman.

FULLER, W. A., 1960. Behaviour and social organization of the wild bison of Wood Buffalo National Park, Canada. *Arctic, 13:* 3-19.

FULLER, W. A., 1962. The biology and management of the bison of Wood Buffalo National Park. *Wildl. Mgmt. Bull., Ottawa Ser. 1(16).*

GOODCHILD, C. & FRANKENBERG, D., 1962. Voluntary running in the golden hamster, *Mesocricetus auratus* (Waterhouse, 1839) infected with *Trichinella spiralis* (Owen, 1835). *Trans. Am. microsc. Soc., 81:* 292-298.

GRAHAM, G. L., 1966. The behavior of beetles, *Tribolium confusum,* parasitized by the larval stage of a chicken tapeworm, *Raillietina cesticillus. Trans. Am. microsc. Soc., 85:* 163.

GRUS, I., 1966. Prilog poznavanju epizooliologije dikrocelioze i drugog prelaznog domacina na terenima srbije. *Acta vet., Beogr., 16:* 249-255.

HAFTORN, S., 1953. Contribution to the food biology of tits especially about storing of surplus food. I. The crested tit (*Parus c. cristatus* L.). *K. norske Vidensk. Selsk. Skr., 4:* 9-122.

HAFTORN, S., 1956a. Contribution to the food biology of tits especially about storing of surplus food. II. The coal-tit (*Parus a. ater* L.). *K. norske Vidensk. Selsk. Skr., 2:* 5-50.

HAFTORN, S., 1956b. Contribution to the food biology of tits especially about storing of surplus food. III. The willow tit (*Parus atricapillus* L.). *K. norske Vidensk. Selsk. Skr., 3:* 5-78.

HAFTORN, S., 1956c. Contribution to the food biology of tits especially about storing of surplus food. IV. A comparative analysis of *Parus atricapillus* L., *P. cristatus* L. and *P. ater* L. *K. norske Vidensk. Selsk. Skr., 4:* 4-54.

HERTING, G. E., 1966. *Effects of behavior on vulnerability of forage organisms to predation by bowfin and spotted gar.* M.Sc. thesis, University of Missouri, Columbia.

HERTING, G. E. & WITT, A., 1967. The role of physical fitness of forage fish in relation to their vulnerability to predation by bowfin (*Amia calva*). *Trans. Am. Fish. Soc., 96:* 427-430.

HIRST, S. M., 1965. Ecological aspects of big game predation. *Fauna Flora.Pretoria, 16:* 3-15.

HOFFMAN, G. L. & HOVME, J., 1958. The experimental histopathology of the "tumor" on the brain of the stickleback caused by *Diplostomum baeri eucaliae* Hoffman and Hundley, 1957 (Trematoda : Strigeoidea). *J. Parasit., 44:* 374-387.

HOHORST, W., 1964. Die rolle der ameisen im entwicklungsgang des lanzettegels (*Dicrocoelium dendriticum*). *Z. Parasitenk., 22:* 105-106.

HOHORST, W. & GRAEFE, G., 1961. Ameisen–obligatorische zwischenwirte des lanzettegels (*Dicrocoelium dendriticum*). *Naturwissenschaften, 48:* 229-230.

HOHORST, W. & LÄMMLER, G., 1962. Experimentelle dicrocoeliose-studien. *Z. Tropenmed. Parasit., 13:* 377-397.

HOLLING, C. S., 1961. Principles of insect predation. *A. Rev. Ent., 6:* 163-182.

HOLLING, C. S., 1965. The functional response of predators to prey density and its role in mimicry and population regulation. *Mem. ent. Soc. Can., 45:* 3-60.

HOLLING, C. S., 1966a. The functional response of invertebrate predators to prey density. *Mem. ent. Soc. Can., 48:* 1-86.

HOLLING, C. S., 1966b. The strategy of building models of complex ecological systems. In Watt (Ed.), *Systems analysis in ecology.* New York: Academic Press.

HORNOCKER, M. G., 1970. An analysis of mountain lion predation upon mule deer and elk in the Idaho primitive area. *Wildl. Monogr., 21:* 1-39.

JENKINS, D., WATSON, A. & MILLER, G. R., 1963. Population studies on red grouse, *Lagopus lagopus scoticus* (Lath.) in north-east Scotland. *J. anim. Ecol., 32:* 317-376.

JENKINS, D., WATSON, A. & MILLER, G. R., 1964. Predation and red grouse populations. *J. appl. Ecol., 1:* 183-195.

JENKINS, T. M., FELDMETH, C. R. & ELLIOTT, G. V., 1970. Feeding of rainbow trout (*Salmo gairdneri*) in relation to abundance of drifting invertebrates in a mountain stream. *J. Fish. Res. Bd. Can., 27:* 2356-2361.

KENNEDY, C. R., 1970. The population biology of helminths of British freshwater fish. In Taylor & Muller (Eds), *Aspects of fish parasitology, Symp. Br. Soc. Parasit., 8:* 145-159.

KLAAS, E. E., 1963. Ecology of the trematode, *Clinostomum marginatum*, and its hosts in eastern Kansas. *Trans. Kans. Acad. Sci., 66:* 519-538.

LAMBERT, T. C. & FARLEY, J., 1968. The effect of parasitism by the trematode *Cryptocotyle lingua* (Creplin) on zonation and winter migration of the common periwinkle, *Littorina littorea* (L.). *Can. J. Zool., 46:* 1139-1147.

LARSON, O. R., 1965. *Diplostomulum* (Trematoda: Strigeoidea) associated with herniations of bullhead lenses. *J. Parasit., 51:* 224-229.

LAUTENSCHLAGER, E. W., 1959. Meningeal tumors of the newt associated with trematode infection of the brain. *Proc. helminth. Soc. Wash., 26:* 11-14.

LEHMANN, U., 1967. Drift und populations dynamik von *Gammarus pulex fossarum* Koch. *Z. Morph. Okol. Tiere, 60:* 227-274.

LEIBY, P. D. & DYER, W. G., 1971. Cyclophillidean tapeworms of wild Carnivora. In Davis & Anderson (Ed.), *Parasitic diseases of wild mammals, pp. 174-234.* Ames: The Iowa State University Press.

LESTER, R. J. G., 1971. The influence of *Schistocephalus* plerocercoids on the respiration of *Gasterosteus* and a possible resulting effect on the behaviour of the fish. *Can. J. Zool., 49:* 361-366.

LEWIS, D. J. & WRIGHT, C. A., 1962. A trematode parasite of *Simulium. Nature, Lond., 193:* 1311-1312.

LINTON, E., 1906. A cestode parasite in the flesh of the butterfish. *Bull. Bur. Fish., Wash., 26:* 111-132.

LOCKE, L., DeWITT, J., MENZIE, C. & KERWIN, J., 1964. A merganser die off associated with larval *Eustrongylides. Avian Diseases, 8:* 420-427.

MacARTHUR, R. H., 1958. Population ecology of some warblers of north-eastern coniferous forests. *Ecology, 39:* 599-619.

MacARTHUR, R. & MacARTHUR, J. W., 1961. On bird species diversity. *Ecology, 42:* 594-598.

MacARTHUR, R. H. & PIANKA, E. R., 1966. On optimal use of a patchy environment. *Am. Nat., 100:* 603-609.

MARGOLIS, L., 1970. Nematode diseases of marine fishes. In Sniezko (Ed.), *A symposium on diseases of fishes and shellfishes. Am. Fish. Soc. Spec. Publ., 5:* 190-208.

MECH, L. D., 1966. The wolves of Isle Royale. *Fauna of the National Parks of the U.S., Fauna Series, 7.* Washington: U.S. Government Printing Office.

MECH, L. D., 1970. *The wolf: The ecology and behavior of an endangered species.* New York: The Natural History Press.

MILLEMAN, R. E. & KNAPP, S. E., 1970. Pathogenicity of the "salmon poisoning trematode", *Nanophyetus salmincola*, to fish. In Sniezko (Ed.), *A symposium on diseases of fishes and shellfishes. Am. Fish. Soc. Spec. Publ., 5:* 209-217.

MOSSMAN, A. S., 1955. *Experimental studies of fitness as measured by vulnerability to predation.* Ph.D. thesis, University of Wisconsin, Madison.

MUELLER, J. F., 1966. Host-parasite relationships as illustrated by the cestode, *Spirometra mansonoides.* In McCauley (Ed.), *Host-parasite relationships. Proceedings of the Twenty-sixth Annual Biology Colloquium pp. 15-58.* Corvallis: Oregon State Univ. Press.

MURIE, A., 1944. The wolves of Mount McKinley. *U.S. Natl. Park Serv., Fauna Ser. 5:* 238pp.

OLNEY, P. J. S., 1964. The autumn and winter feeding biology of certain sympatric ducks. *Trans. Sixth Int. Union Game Biol.:* 309-322.

OLSON, R. E., 1968. *The life cycle of* Cotylurus erraticus *(Rudolphi, 1809) Szidat, 1928 (Trematoda: Strigeidae) and the effect of the metacercariae on rainbow trout* (Salmo gairdneri). Ph.D. thesis, Montana State University, Bozeman.

ORR, T. S. C., 1966. Spawning behaviour of rudd, *Scardinius erythrophthalmus,* infested with plerocercoids of *Ligula intestinalis. Nature, Lond., 212:* 736.

PERRET, N. G., 1962. *The spring and summer foods of the common mallard (Anas platyrhynchos platyrhynchos* L.) *in south central Manitoba. M.Sc. thesis, University of British Columbia, Vancouver.*

PIELOWSKI, Z., 1959. Investigation of the predator (hawk)—victim (pigeon) system. *Bull. Acad. pol. Sci. Sér. Sci. Biol., 7:* 401-403.

PODESTA, R. B. & HOLMES, J. C., 1970. The life cycles of three polymorphids (Acanthocephala) occurring as juveniles in *Hyalella azteca* (Amphipoda) at Cooking Lake, Alberta. *J. Parasit., 56:* 1118-1123.

POPHAM, E. J., 1942. Further experimental studies of the selective action of predators. *Proc. zool. Soc. Lond., 112:* 105-117.

RACTLIFFE, L. H., TAYLOR, H. M., WHITLOCK, J. H. & LYNN, W. R., 1969. Systems analysis of a host-parasite interaction. *Parasitology, 59:* 649-661.

RAUSCH, R. L., 1952. Hydatid disease in boreal regions. *Arctic, 5:* 157-174.

RITCEY, R. W. & EDWARDS, R. Y., 1958. Parasites and diseases of the Wells Gray moose herd. *J. Mammal., 39:* 139-145.

ROTHSCHILD, M., 1940. *Cercaria pricei,* a new trematode, with remarks on the specific characters of the "Prima" group of Xiphidiocercariae. *J. Wash. Acad. Sci., 30:* 437-448.

ROTHSCHILD, M., 1962. Changes in behavior in the intermediate hosts of trematodes. *Nature, Lond., 193:* 1312-1313.

RUDEBECK, G., 1950; 1951. The choice of prey and modes of hunting of predatory birds with special reference to their selective effect. *Oikos, 2:* 65-88; *3:* 200-231.

SCHAD, G. A., 1966. Immunity, competition, and natural regulation of helminth populations. *Am. Nat., 100:* 359-364.

SCHALLER, G. B., 1967. *The deer and the tiger.* Chicago: University Chicago Press.

SCHOENER, T. W., 1969. Models of optimal size for solitary predators. *Am. Nat., 103:* 277-313.

SILLMAN, E. I., 1957. A note on the effect of parasite burden on the activity of fish. *J. Parasit., 43:* 100.

SINDERMANN, C. J., 1960. Ecological studies of marine dermatitis-producing schistosome larvae in northern New England. *Ecology, 41:* 678-684.

SINDERMANN, C. J., 1970. *Principal diseases of marine fish and shellfish.* New York: Academic Press.

SINDERMANN, C. J. & FARRIN, A. E., 1962. Ecological studies of *Cryptocotyle lingua* (Trematoda: Heterophyidae) whose larvae cause "pigment spots" of marine fish. *Ecology, 43:* 69-75.

SLOBODKIN, L. B., 1968. How to be a predator. *Am. Zool., 8:* 43-51.

SMYTH, J. D., 1964. The biology of the hydatid organisms. *Adv. Parasitol., 2:* 169-219.

SMYTH, J. D., 1969. The biology of the hydatid organisms; literature review—1964-68. *Adv. Parasitol., 7:* 327-347.

SMYTH, J. D. & HEATH, D. D., 1970. Pathogenesis of larval cestodes in mammals. *Helminth. Abstr., 39:* 1-22.

SOGANDARES-BERNAL, F. & LUMSDEN, R. D., 1964. The heterophyid trematode *Ascocotyle* (A.) *leighi* Burton, 1956, from the hearts of certain poeciliid and cyprinodont fishes. *Z. Parasitenk. 24:* 1-12.

SPALL, R. D. & SUMMERFELT, R. C., 1970. Life cycle of the white grub, *Posthodiplostomum minimum* (MacCallum, 1921: Trematoda, Diplostomatidae), and observations on host-parasite relationships of the metacercariae in fish. In Sniezko (Ed.), *A symposium on diseases of fishes and shellfishes. Am. Fish Soc. Spec. Publ., 5:* 218-230.

SPARKS, A. K. & CHEW, K. K., 1966. Gross infestation of the littleneck clam, *Venerupis staminea,* with a larval cestode (*Echeneibothrium* spp.). *J. Invert. Path., 8:* 413-416.

SPRENT, J. F. A., 1955a. On the invasion of the central nervous system by nematodes. I. The incidence and pathological significance of nematodes in the central nervous system. *Parasitology, 45:* 31-40.

SPRENT, J. F. A., 1955b. On the invasion of the central nervous system by nematodes. II. Invasion of the nervous system in ascariasis. *Parasitology, 45:* 41-55.

SPRENT, J. F. A., 1962. The evolution of the Ascaridoidea. *J. Parasit., 48:* 818-824.

SUGDEN, L. G., 1965. Food and food energy requirements of young ducklings. *Annual progress report, 1964-65. Canadian Wildlife Service (unpubl. report).*

SUGDEN, L. G., 1969. *Foods, food selection and energy requirements of wild ducklings in southern Alberta.* Ph.D. thesis, Utah State University, Logan.

SWENNEN, C., 1969. Crawling-tracks of trematode infected *Macoma balthica* (L.). *Neth. J. Sea Res., 4:* 376-379.

SZIDAT, L., 1969. Structure, development, and behavior of new strigeatoid metacercariae from subtropical fishes of South America. *J. Fish. Res. Bd. Can., 26:* 753-786.

SZIDAT, L. & NANI, A., 1951. Diplostomiasis cerebralis del pejerrey. *Rev. Inst. nac. Invest. Cienc. nat.,* *1:* 323-384.

TINBERGEN, L., 1960. The natural control of insects in pine woods. I. Factors influencing the intensity of predation by songbirds. *Archs néerl. Zool., 13:* 265-343.

TINER, J. D., 1953a. The migration, distribution in the brain, and growth of ascarid larvae in rodents. *J. infect. Dis., 92:* 105-113.

TINER, J. D., 1953b. Fatalities in rodents caused by larval *Ascaris* in the central nervous system. *J. Mammal., 34:* 153-167.

TINER, J. D., 1954. The fraction of *Peromyscus leucopus* fatalities caused by raccoon ascarid larvae. *J. Mammal., 35:* 589-592.

TULLOCK, G., 1971. The coal-tit as a careful shopper. *Am. Nat., 105:* 77-79.

ULMER, M. J., 1971. Site-finding behavior in helminths in intermediate and definitive hosts. In Fallis (Ed.), *Ecology and physiology of parasites.* Toronto: University of Toronto Press.

WALKEY, M. & MEAKINS, R. H., 1970. An attempt to balance the energy budget of a host-parasite system. *J. Fish. Biol., 2:* 361-372.

WARNER, R. W. & KATKANSKY, S. C., 1969. Infestation of the clam *Protothaca staminea* by two species of tetraphyllidian cestodes (*Echeneibothrium* spp.). *J. Invert. Path., 13:* 129-133.

WEIDMANN, U., 1956. Verhaltangsstudien an der stockente. I. Das aktionssystem. *Z. Tierpsychol., 13:* 208-268.

WESENBURG-LUND, C., 1931. Contributions to the development of the Trematoda Digenea. I. The biology of *Leucochloridium paradoxum. K. dansk. Vidensk. Selsk. Skr. 9th Series, 4:* 89-142.

WHEELER, W. M., 1910. *Ants, their structure, development and behavior.* New York: Columbia University Press.

WHITTAKER, R. H. & FEENEY, P. P., 1971. Allelochemics: chemical interactions between species, *Science, N.Y., 171:* 757-769.

WILLIAMS, H. H., 1967. Helminth diseases of fish. *Helminth Abstr., 36:* 261-295.

WRIGHT, B. S., 1960. Predation on big game in East Africa. *J. Wildl. Mgmt, 24:* 1-15.

WRIGHT, C. A., 1966. The pathogenesis of helminths in the Mollusca. *Helminth. Abstr., 35:* 207-224.

YAMAGUTI, S., 1958. *Systema helminthum,* I. *Digenetic trematodes.* New York: Interscience Publ.

DISCUSSION

R. M. Cable

Did you use in your experiments any gammarids with multiple infections? Is there any change in the specific gravity of infected *Gammarus*?

J. C. Holmes

No. The cystacanth is a large one, and multiple infections may produce other deleterious effects on the gammarid. In the lab, gammarids with multiple infections appear weaker and die earlier (although we haven't proven that) than those with only a single cystacanth. We wanted to avoid possible complications of that nature. In the wild, we find very few gammarids with more than a single mature cystacanth.

We have not measured the specific gravity of infected and uninfected gammarids, but the behaviour of infected gammarids in aquaria lighted from below (covered in the written paper, but not that delivered orally) suggests that the altered behaviour of the infected gammarids is unrelated to any such change.

S. B. Kendall

Dicrocoelium dendriticum similarly disturbs the behaviour of the ant intermediate host and leads to selective predation. Has Dr Holmes any views on the evolutionary significance of the phenomenon?

J. C. Holmes

The phenomenon would appear to have evolutionary significance for the parasite and for the intermediate host. Our major point is that anything the parasite does which alters the behaviour of infected intermediate hosts so as to result in selective predation by a suitable host, has obvious survival value and will be selected for.

We have also suggested that altered behaviour of infected individuals may have survival value for the intermediate host population. Neither of us has seen ants infected with *D. dendriticum,* but we have seen those infected with the related *Brachylecithum mosquensis.* The infected ants are obviously poorer foragers than uninfected ants. (It would be interesting to determine if they are a net energy loss to the ant colony.) Assuming that some ants are going to be lost to predators, that loss would be less expensive to the colony if it involved the less productive individuals. If so, responses of the less productive, infected individuals which altered their behaviour making them more susceptible to predation should also be selected for, particularly if the search image formed by the predator tended to protect the more productive members of the colony. This appears to be the case with ants infected with *B. mosquensis* (see Carney, 1969).

R. Muller

In the example quoted by Dr Kendall the change in behaviour of the ant is due to the presence of the metacercaria of *Dicrocoelium* in the host's brain. Have you any idea of the mechanism by which the presence of a *Polymorphus* larva alters the behaviour of *Gammarus?*

J. C. Holmes

Hirnwürmer are associated with the changed behaviour of ants infected with either *D. dendriticum* or *B. mosquensis.* Carney's observations on the latter indicate that the metacercariae are not always in the brain, but may be adjacent to it. The possibility of a chemical mediator, an allomone, cannot be discarded. Mr Bethel is investigating the possibility of an allomone in the *Polymorphus paradoxus—Gammarus* system.

R. H. Meakins

The presence of large numbers of small plerocercoids of *Schistocephalus solidus* disturbs the behaviour of the fish host so that it is liable to predation. These small larvae are, however, unable to mature in the definitive host so that there is a tremendous loss to the system.

J. C. Holmes

All parasite systems that I am acquainted with involve tremendous losses or wastage, particularly where transmission rates are high, as in some of your studies with *S. solidus.* Losses such as you describe can be tolerated by the parasite if the system is generally efficient. In addition, the possible advantages of selective predation to the stickleback population would be independent of maturation of the parasite. Thus, selective predation on sticklebacks with large numbers of small plerocercoids might not be harmful to the *Schistocephalus* population, but advantageous to the stickleback population.

D. Davenport

Have experiments been conducted to determine whether there is, in fact, a higher survival rate among non-parasitized gammarids when parasitized individuals are present than when parasitized individuals are *not* present?

J. C. Holmes

No, not yet. We first formulated that hypothesis while writing this paper. It is an intriguing possibility, one well worth pursuing, isn't it?

N. A. Croll

Are there not selection pressures acting on behalf of the definitive host which may have endowed it with the ability to avoid infected intermediate hosts? Are definitive hosts always idiots?

Plate 1

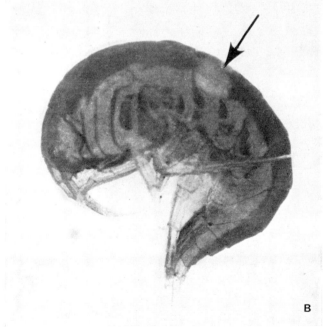

J. C. HOLMES AND W. M. BETHEL

Plate 2

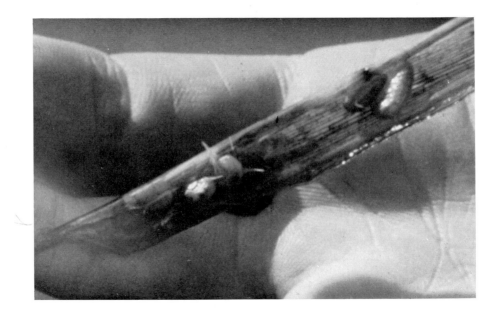

J. C. HOLMES and W. M. BETHEL

J. C. Holmes

Definitive hosts are not idiots. Quite the contrary. The systems we describe are absolutely dependent upon the ability of definitive hosts to feed so as to maximize their energy intake per unit energy or time expended. The system involves a trade-off, with acquisition of parasites being traded for easier capture of prey. So long as the parasites acquired do not cost the definitive host population more energy than it saves through easier predation, the system would be advantageous to the predator. It must be emphasized that it is the population's energy balance that is involved; individual hosts could occasionally lose out.

In the *Polymorphus–Gammarus*–mallard system that we have studied, there is very little apparent pathology in the mallards. The metabolism of the worms and the metabolic disturbances they produce (see Crompton, 1970) involve some metabolic drain, but indications so far are that the drain is nearly negligible. If so, in that system, at least, there should be very little selection pressure towards mechanisms to avoid infected intermediate hosts. Where the parasites produce more damage, the system is more complex, and may involve avoiding infected intermediate hosts. Of course, vertebrate definitive hosts do have other methods of reducing the energy drain to the parasites they ingest, such as the development of an effective immune response. The marginal energy cost of such a response would probably usually be less than the energy saved by the more effective predatory system.

EXPLANATION OF PLATES

PLATE 1

A. Crawling tracks of trematode-infected *Macoma balthica* on sandy tidal flat (from Swennen, 1969; courtesy of E. J. Brill, Leiden).
B. A cystacanth of *Polymorphus paradoxus* (arrow) in the haemocoele of *Gammarus lacustris*.

PLATE 2

Gammarus lacustris infected with *Polymorphus paradoxus* clinging onto a reed.

Influence of the behaviour of amphibians on helminth life-cycles

C. COMBES

Centre Universitaire de Perpignan, France

From an ecological point of view there are two ways in which platyhelminths infect vertebrates, either by direct entry or indirectly via a link in the food chain leading to the vertebrate.

The part played by the behaviour of the definitive host in these processes has been studied in the Amphibia. The life-cycles of platyhelminths parasitic in amphibians have been classified into five groups according to the type of infestation (direct or indirect) and the type of food-chain (wholly aquatic, part aquatic-part terrestrial, wholly terrestrial). The parasites usually start their cycles in an aquatic environment and some means of conveyance is necessary to effect the transfer from an aquatic to a terrestrial ecosystem. This may be through a vector such as an insect with an aquatic larva or through the behaviour of the amphibian host which may, for instance, have a phase of feeding in water. Some of the life-cycles however are either completely aquatic or entirely terrestrial.

The life-cycle of a given platyhelminth normally follows a particular pattern (the habitual mode) but a change in environmental conditions or in the species of definitive host may modify the habitual mode and give alternative patterns.

Quantitative studies of the influence of amphibian behaviour on some of the cycles have been made.

A list of the better known life-cycles of platyhelminths parasitic in amphibians is given; it is based as far as possible on the habitual modes of transmission which have been described.

CONTENTS

INTRODUCTION

In the infestation of vertebrates by parasitic platyhelminths, one can observe two main processes *in which the host behaviour intervenes*:

(1) a process in which the platyhelminth is transmitted to the definitive host;

(2) a process in which the platyhelminth, which is a fundamentally aquatic animal, can leave one ecosystem for another.

Process in which the platyhelminth is transmitted
to the definitive host

Two cases are noticeable:

(1) the parasite, through its own behaviour, penetrates directly the vertebrate (direct infestation);

(2) the parasite enters into a food chain leading eventually to the vertebrate, in order to be swallowed (indirect infestation).

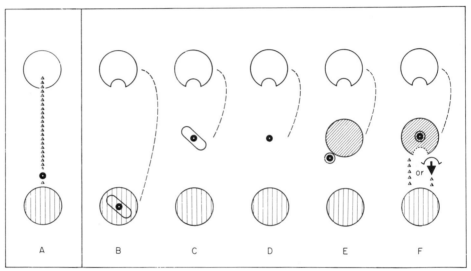

Figure 1. Infestation of the definitive host by Digenea: direct infestation (A) and indirect infestation (B to F).
Vertical shading, molluscs; diagonal shading, vector; white, definitive host; black circle, infesting stage of the parasite.

Figure 1 shows direct infestation (A) and different ways of indirect infestation (B to F) in the case of Digenea:

A, the cercaria penetrates directly the definitive host: e.g.: *Schistosoma haematobium*, a parasite of man (in all cases of direct infestation the phenomenon takes place compulsorily in an aquatic environment);

B, predation of the mollusc or part of the mollusc in which the cercariae are lying in wait: e.g.: *Leucochloridium paradoxum*, a parasite of birds;

C, predation of the sporocyst escaped from the mollusc and containing the cercariae: e.g.: *Plagioporus sinitzini*, a parasite of fresh-water fish;

D, predation of the free-swimming cercaria: e.g.: *Azygia lucii*, a parasite of fresh-water fish;

E, predation of the cercaria encysted *on* an edible organism: e.g.: *Fasciola hepatica,* a parasite of herbivorous mammals;

F, predation of the cercaria encysted *inside* an edible organism: e.g.: *Paragonimus westermanni,* a parasite of man.

In the last two examples, a vector (or intermediate host) appears between the mollusc and the vertebrate; this intermediate host can in its turn receive the parasite either directly or as part of a food chain.

With Monogenea, only direct infestation exists; with Cestoda, this direct infestation is never observed.

Process in which the platyhelminth can leave one ecosystem for another

Continuing the example of the Digenea, we shall classify the processes of transport of the parasite from an aquatic biocenosis to a terrestrial one.

In the case of direct infestation, one may find the three following cases:

(1) mollusc and definitive host both aquatic: no transport;

(2) mollusc aquatic; definitive host aquatic in its larval stage, then terrestrial;

(3) mollusc aquatic; definitive host terrestrial, but temporarily sojourning in water.

In indirect infestation (Fig. 2) the processes are more varied as in the following examples:

A, mollusc, vector and definitive host all aquatic; no change of environment;

B, mollusc and vector aquatic; the definitive host swallows the vector during its larval stage (aquatic) then leaves for another ecosystem;

C, mollusc and vector aquatic; the definitive host (terrestrial) gets its food from an aquatic ecostystem;

D, mollusc aquatic; the vector is aquatic during its larval stage, then carries the parasite into a terrestrial environment;

E, mollusc, vector and definitive host all terrestrial; no change of environment;

F, mollusc and definitive host terrestrial; no vector (the metacercaria encysts on the spot); no change of environment.

Processes similar on the whole, but different of course in their forms can be observed with Monogenea and Cestoda.

From the preceding remarks, two aspects of the behaviour of vertebrates emerge: (1) one regarding feeding; (2) the other regarding water.

These two aspects of behaviour, in fact closely interrelated (for the type of prey swallowed is dependent on the environment) are indeed of the highest importance with amphibians of which the majority live on the fringe of aquatic and terrestrial environments. We shall consider the influence of the behaviour of amphibians on the life-cycles of platyhelminths, first qualitatively, then quantitatively.

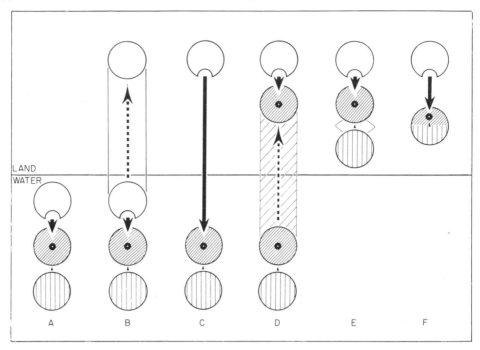

Figure 2. Change of ecosystem of the Digenea in the case of indirect infestation (see text).

INFLUENCE OF THE BEHAVIOUR OF AMPHIBIANS ON THE CYCLES

Qualitative study

From the behavioural aspects of parasite transmission, we can classify the modes of infestation in the following manner; five groups may be distinguished:

(1) direct infestation; no predatory behaviour of the amphibian*; the infestation takes place in the tadpole stage;

(2) direct infestation; no predatory behaviour of the amphibian; the infestation takes place in the adult stage;

(3) indirect infestation; a predatory behaviour of the amphibian†; predation of an aquatic animal;

(4) indirect infestation; a predatory behaviour of the amphibian; predation of a terrestrial animal having an aquatic larva;

(5) indirect infestation; a predatory behaviour of the amphibian; predation of an entirely terrestrial animal.

In these five groups of infestation modes, the study of the kinds of organisms which play a part in the cycle allows us to describe several different types. I want to emphasize that the following classification is that of the modes of infestation and not of the life-cycles themselves (these are mentioned only as examples).

* Whenever the amphibian is not a definitive host but a vector, a predatory behaviour is necessary in the definitive host.

† Whenever the amphibian is not a definitive host but a vector, the food chain extends to the definitive host.

Group A. Direct infestation of the tadpole, Fig. 3

In this mode of infestation, the parasite having infected the tadpole in water is carried out to a terrestrial environment with the metamorphosis of the amphibian (in some cases the parasite dies at the metamorphosis).

This mode of infestation is frequent with the Monogenea Polystomatidae (type A1; ex: *Polystoma integerrimum*) and it occurs also with the rare species of Dactylogyridae parasitic on the gills of amphibian larvae (in this case the parasite does not survive in the adult amphibian).

This mode is rare with the Digenea of amphibians (type A2; ex: *Opisthioglyphe rastellus*), in which the process is complicated: the cercaria after having actively penetrated the body, encysts, then excysts and makes for its eventual microbiotope; on the other hand, this mode of infestation is very frequent in the cycles of platyhelminths parasitic in batrachophagous vertebrates; the amphibian takes the parasite out of water (as a metacercaria) on to the land; here one link of a food chain (type A3; ex: *Procyotrema marsupiformis*) or two (type A4; ex: *Strigea elegans*) will follow. But we must notice that, when the carnivorous animal is able to eat the tadpoles in the water, the metamorphosis of the amphibian is no longer necessary (type A5; ex: *Leptophallus nigrovenosus*).

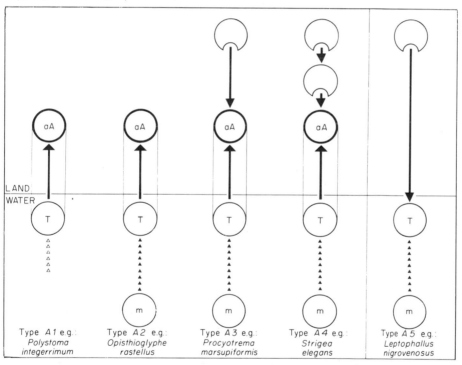

Figure 3. Mode of infestation of platyhelminths parasitizing amphibians: Group A. aA, Adult amphibian; m, mollusc; T, tadpole.

Group B. Direct infestation of the adult, Fig. 4

In this mode of infestation, the parasite infects the adult in an aquatic environment, either because the amphibian stays in water all its life long, or because it sojourns temporarily in it.

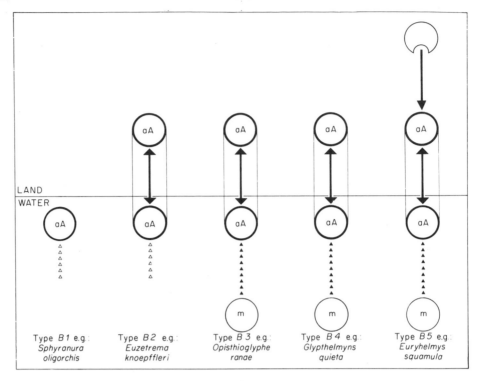

Figure 4. Mode of infestation of platyhelminths parasitizing amphibians: Group B. aA, Adult amphibian; m, mollusc.

This mode exists both with Monogenea and Digenea.

With the monogenean parasites of amphibians living all the time in water, the ciliated larva infects naturally the adult as well as the tadpole (type B1; ex: *Sphyranura oligorchis*); when the adult amphibian is aquatic only occasionally, the infestation takes place while bathing (type B2; ex: *Euzetrema knoepffleri*); it is particularly noteworthy that, with *Protopolystoma xenopi,* a parasite of a toad (*Xenopus*) which is aquatic most of the time, the larva enters the bladder by the cloaca, in contrast to what has been observed with *Polystoma integerrimum.* With *Gyrdicotylus gallieni,* a parasite of the oral cavity of *Xenopus,* the larva is not ciliated and creeps on the bottom of the ponds until it meets the host and then it reaches the microbiotope in which it can survive on the rare occasions when the toad goes out of water (Tinsley, pers. comm.).

With the Digenea, the type B3 (ex: *Opisthioglyphe ranae*) seems to be rather rare: the cercaria penetrates the body during a bath, encysts, excysts and settles in the gut. More frequently, the cercaria encysts in the skin and the metacercaria is ingested with the skin during sloughing (type B4; ex: *Glypthelmyns quieta*). We see here the first trophic behaviour of the amphibian, transitional with the following groups (we keep this type in the B group because the parasite, as in the other types of this group, infects directly the host in which it will grow to adulthood). Of course, the amphibian may be acting as an intermediate host for a batrachophagous vertebrate (type B5; ex: *Euryhelmys squamula*).

Group C. Predatory behaviour of the amphibian, aquatic prey, Fig. 5

Theoretically, the cycles of this group should be limited to Urodela for the tadpoles of Anura are herbivorous and adult Anura do not feed much under water. However, some Digenea and Cestoda of Anura as well as Urodela have this mode of infestation.

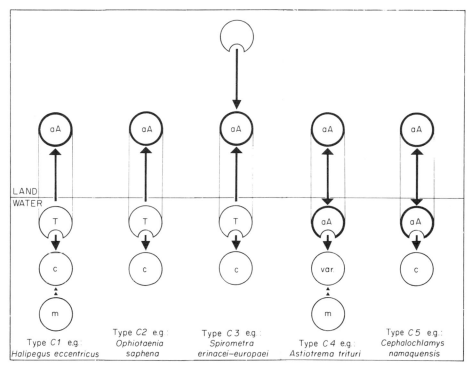

Figure 5. Mode of infestation of platyhelminths parasitizing amphibians: Group C. aA, Adult amphibian; c, crustacea; i, insect; m, mollusc; T, tadpole; var., variable.

The tadpoles of Anura can swallow copepods carrying metacercariae (type C1; ex: *Halipegus eccentricus*) or infesting procercoids (type C2; ex: *Ophiotaenia saphena*). The amphibian may be acting as an intermediate host for the Cestoda of a batrachophagous animal (type C3; ex: *Spirotrema erinacei-europaei*) and this kind of cycle is probably the most normal with the Pseudophyllidea for they have both a procercoid and a plerocercoid. There are other cases when the adult amphibian ingests aquatic vectors carrying metacercariae (type C4; ex: *Astiotrema trituri*) or procercoids (type C5; ex: *Cephalochlamys namaquensis*). Usually, an adult urodele swallows Cladocera, molluscs or even fishes (this is the case with the giant salamander of Japan); sometimes however, it is an anuran, the most obvious example being that of *Xenopus* which eats Copepoda (the case of *Rana nigromaculata* which is supposed to eat shrimps (Shibue, 1953) is a bit more difficult to account for). We must remark that, with some amphibians which are most of the time in the water, the cycles closely resemble the cycles in fish and the parasite is not really removed from water; in these cycles the larva and the adult play frequently the same part.

Group D. Predatory behaviour of the amphibian, terrestrial prey
but with aquatic larva, Fig. 6

It is well known that the food of adult amphibians consists mainly of insects with the occasional addition of some other Arthropoda, molluscs and smaller vertebrates (including their own young). For the transmission of platyhelminths (which are fundamentally aquatic animals) organisms having an ecology

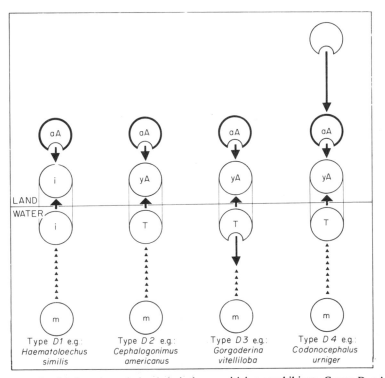

Figure 6. Mode of infestation of platyhelminths parasitizing amphibians: Group D. aA, Adult amphibian; i, insect; m, mollusc; T, tadpole; yA, young amphibian.

successively aquatic then terrestrial are of great interest for they provide a new means of taking parasites out of water. Insects with aquatic larvae are very usual prey of amphibians which live near the water side, and they act as the vectors of many Digenea (type D1; ex: *Haematoloechus similis*). The predatory behaviour of adult amphibians towards their own young places the latter in the same position as the insects (type D2; ex: *Cephalogonimus americanus*). It may be mentioned here that in some cases, the predatory behaviour of the tadpole towards the cercaria is a variation on the pattern (type D3; ex: *Gorgoderina vitelliloba*). As in the preceding groups, the amphibian may be only a stage in the unfolding of the cycle (type D4; ex: *Codonocephalus urniger*).

Group E. Predatory behaviour of the amphibian, terrestrial prey, Fig. 7

Whereas in the preceding groups the parasite generally has to be taken out of fresh water to reach eventually the adult amphibian, there are a few cases in which the whole cycle takes place terrestrially. This cycle may include two invertebrates (type E1; ex: *Brachycoelium mesorchium*) or only the mollusc

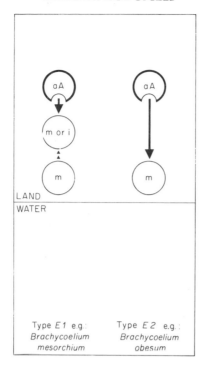

Figure 7. Mode of infestation of platyhelminths parasitizing amphibians: Group E. aA, Adult amphibian; i, insect; m, mollusc.

(type E2; ex: *Brachycoelium obesum*). These terrestrial invertebrates are swallowed by the amphibians.

Several important remarks must be made:

(1) It is essential never to forget that the lines above describe a classification of the means of transmission and under no circumstances the cycles themselves (these are referred to only as examples). In fact, with most cycles, although the habitual mode is the one described, other modes nearly always play a part. For instance, in the cycle of *Opisthioglyphe rastellus* direct infestation can also take place with the adult and not with the tadpole; in the cycle of *Opisthioglyphe ranae*, direct infestation can also take place with the tadpole and not with the adult; it is only recently that Grabda-Kazubska (1969) has shown that infestation of the tadpole is preeminent with *Opisthioglyphe rastellus* and infestation of the adult is preeminent with *Opisthioglyphe ranae*. Similarly, in the cycles of the type A5, the metacercariae may of course be swallowed when they are in the adult amphibian and so we have an infestation of the type A3: it all depends on the behaviour of the batrachophagous animal (whether a tadpole eater or a frog eater).

(2) In a study of the behavioural aspects of parasite transmission, I think it is necessary to find out the habitual mode among the possible modes for various hosts and countries. By analogy with the definition I gave (1968 : 38-39) of the different kinds of hosts, I think we can speak for each cycle of the possible modes, the habitual mode, the accidental modes. Of

course, the habitual mode may differ in relation with the host behaviour; for
instance in the cycle of a digenean having as a vector an insect with aquatic
larvae, we can suppose that:

(i) a urodele will be infected by eating the insect larvae in water (infestation
of the type C4);

(ii) an anuran will be infected by eating the insect imago (infestation of the
type D1).

(3) Very often the modes of infestation used by parasites depend on the
behaviour of the hosts rather than on their taxonomic position; for instance, in
France, *Rana temporaria* and *Bufo bufo* (mainly terrestrial Anura) are
parasitized by *Opisthioglyphe rastellus* and *Haplometra cylindracea* (direct
infestation taking place mainly in the *tadpoles*), but *Rana esculenta* and *Rana
ridibunda* (mainly aquatic ones) are parasitized by *Opisthioglyphe ranae* (direct
infestation taking place mainly in the *adult*).

(4) There is nearly always a close relation between the behaviour of the
hosts and that of the parasites so both must be studied together. Sometimes,
this relation seems to be the result of a slow evolution. I will choose as an
example the cycle of Polystomatidae (Fig. 8): parasitism of terrestrial animals
is easy with Digenea and with Cestoda which are endoparasites, but it is
difficult with Monogenea which cannot possibly survive out of an aquatic

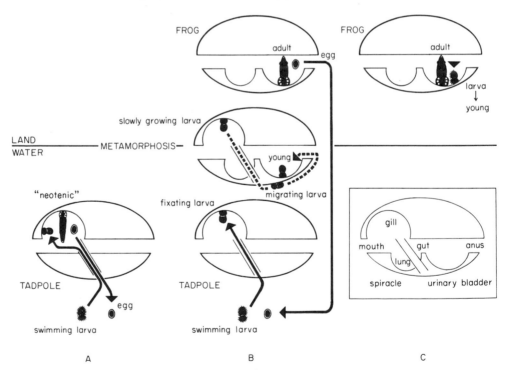

Figure 8. Evolution of the cycle of the Polystomatidae (Monogenea).
   A. Intervention of a so-called "neotenic" form, producing eggs on the gills of the tadpoles
and which can be repeated in succession. B. Infestation of the tadpole followed with a change
of microbiotope at the time of the metamorphosis of the Amphibian. C. Internal cycle in the
bladder of the adult frog.

environment. The larvae of polystomatids first attach to the gills of tadpoles then migrate at the time of metamorphosis to the bladder which provides an aquatic microenvironment. However, with certain species, the same parasite may lay eggs on the gills of the tadpole (the so-called neotenic form) and an internal cycle (allowing, it is true, no dispersal) may appear in the bladder of the adult frog. Perhaps we see here the adaptation of the cycle to the host behaviour in the course of evolution?

Quantitative study

The nutritional habits of the definitive host and its haunting of water may influence the degree of infestation by platyhelminths in a given biotope.

As regards amphibians, accurate information on this topic is not abundant.

I shall select for comparative analysis the influence of the behaviour of two amphibians susceptible to the same parasites; then I shall study the influence of seasonal behaviour of a given amphibian.

Comparative influence of the behaviour of two amphibians

Plasota (1969) has shown (Fig. 9A) that in the Kampinos National Park (Poland), the two Digenea to be found both in *Rana esculenta* L. (very aquatic)

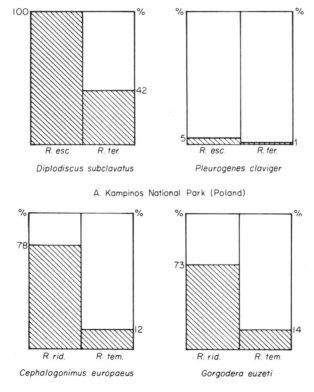

A. Kampinos National Park (Poland)

B. Eastern Pyrenees (Spain)

Figure 9. Comparison of the frequencies of infestation in aquatic and terrestrial frogs (see text). *R. esc., Rana esculenta; R. terr., Rana terrestris; R. rid., Rana ridibunda; R. temp., Rana temporaria.*

and *Rana terrestris* Andrz (mainly terrestrial), do not occur with the same frequency: the aquatic frog is always the more infected. This can be easily accounted for by the modes of infestation: *Diplodiscus subclavatus* is transmitted according to the mode B4, *Pleurogenes claviger* according to the mode D1.

Similarly, Combes & Gerbeaux (1970) have shown (Fig. 9B) in a place in Northern Spain where *Rana ridibunda* Goeze (very aquatic) and *Rana temporaria* L. (mainly terrestrial) live together, the former is the more heavily infected by *Cephalogonimus europaeus* (mode D2) and *Gorgodera euzeti* (mode D1).

It is probable that in certain cases, one amphibian would be a possible host for a given parasite but its behaviour is unfavourable to the transmission, and the frequency becomes nil.

Influence of seasonal behaviour

I have already shown (Combes, 1968 : 93) that by the side of a given lake in the Pyrenees, one can notice every year between the spring and the autumn a very important rise, then a decrease, of the frequency of *Gorgodera euzeti* in common frogs; other parasites have their own variations (Fig. 10).

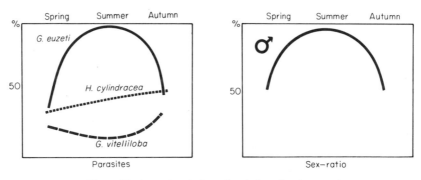

Figure 10. Annual variation of an infestation (see text).

This variation in frequency can be accounted for in the following manner: *Gorgodera euzeti,* transmitted by the imago of *Sialis* (Insecta, Megaloptera) whose larvae live in the lake, colonize heavily in summer those frogs which live on the shore. An important fraction of the frog population are then some distance away from the lake and ingest far more Orthoptera than Megaloptera and consequently are far less infected (which is confirmed by the statistics).

The return of these migratory frogs to the side of the lake in the autumn causes the apparent frequency of the parasites to drop. Study of the sex-ratio along the shore has shown that it is mainly the females that leave the aquatic environment. As regards the other parasites, this migration does not interfere with their frequencies, the infestation not being related to the haunting of the lake shore in the summer.

We have here an interesting case in which the different behaviour of the sexes of the amphibian host affects the course of the parasitic cycle.

CONCLUSION

Apart from the elements of behaviour of the amphibians which we have considered, others may have an influence on the course of certain cycles. The following examples can be mentioned:

(1) The gathering of tadpoles near the shores of lakes (group effect), near to the place where the eggs were laid, is bound to be favourable to the infestation of these tadpoles by the larvae of *Polystoma* species whose eggs are laid at the same time as those of the amphibian.

(2) The type of prey may be significant; for instance, toads are generally far more myrmecophagous than frogs and the latter far more malacophagous than toads; such differences may favour one life cycle more than another.

(3) The fact that an amphibian may have a diurnal or nocturnal activity cycle makes it play a different part in the transmission of helminths to batrachophagous animals, for instance diurnal or nocturnal raptores.

(4) The effect of camouflage (stillness, homochromy) may be a delaying factor in the completion of certain cycles in which the amphibian has to be swallowed.

(5) The alarm calls and warning signals described in certain species sometimes cause a whole population of frogs to run for shelter into water, where certain direct infestations can then take place.

(6) Lastly, the migrations, up to several dozens of miles, observed among certain toads, may allow the transportation of a parasite to entirely new areas.

Unfortunately, it would be presumptuous at this stage to attempt a precise analysis of the influence of these factors on the cycles, for data are not yet adequate. One may suppose, however, that these factors are minor ones and that, among amphibians, it is their behaviour in relation to water and nutritional habits which influence above all the completion of the cycles of the platyhelminths.

REFERENCES

ABDEL AZIM, M., 1935. On the life history of *Lepoderma ramlianum* Looss, 1896 and its development from a xiphidiocercaria. *J. Parasit., 21*(5): 365-368.

ALVEY, C. H., 1936. The morphology and development of the monogenetic trematode *Sphyranura oligorchis* (Alvey, 1933) and the description of *Sphyranura polyorchis* n. sp. *Parasitology, 28*(2): 229-253.

ANDERSON, G. A., 1964. Digenetic trematodes of *Ascaphus truei* in western Oregon. *Diss. Abstr., 25*(6): 3752-3753.

ANDERSON, G. A. & PRATT, I., 1965. Cercaria and first intermediate host of *Euryhelmys squamula. J. Parasit., 51*(1): 13-15.

ANDERSON, G. A., MARTIN, G. W. & PRATT, I., 1966. The life cycle of the trematode *Cephalouterina dicamptodoni* Senger et Macy, 1953. *J. Parasit., 52*(4): 704-706

ANDERSON, G. A., SCHELL, S. C. & PRATT, I., 1965. The life cycle of *Bunoderella metteri* (Allocreadiidae : Bunoderinae), a trematode parasite of *Ascaphus truei. J. Parasit., 51*(4): 579-582.

BITTNER, H., 1925. Ein Beitrag zur Uebertragung und zur Morphologie von *Echinoparyphium recurvatum. Berl. tierärztl. Wschr., 41*(6): 82-86.

BOSMA, N. J., 1934. The life history of the new trematode *Alaria mustelae* Bosma, 1931. *Trans. Am. microsc. Soc., 53*(2): 116-153.

BUTTNER, A., 1952. Nouvelle démonstration d'un cycle abrégé chez *Ratzia joyeuxi* (Trematoda, Opisthorchiidae). *C. r. hebd. Séanc. Acad. Sci., Paris, 234*(6): 673-675.

COMBES, C., 1968. Biologie, Ecologie des cycles et Biogéographie de Digènes et Monogènes d'Amphibiens dans l'Est des Pyrénées. *Mém. Mus. natn. Hist. nat., Paris*, A *51*(1): 1-195.

COMBES, C. & GERBEAUX, M. T., 1970. Recherches éco-parasitologiques sur l'helminthofaune de *Rana ridibunda perezi* (Amphibien Anoure) dans l'Est des Pyrénées. *Vie Milieu, 21*(1C): 121-158.

COMBES, C. & KNOEPFFLER, L. P., 1970. Les Amphibiens et le milieu. *Vie Milieu, 21*(1C): 159-174.

CHENG, T. C., 1960. The life history of *Brachycoelium obesum* Nicholl, 1914, with a discussion of the systematic status of the trematode family Brachycoeliidae Johnston, 1912. *J. Parasit., 46*(4): 464-474.

CRAWFORD, W. W., 1939. Studies on the life history of Colorado trematodes. *J. Parasit., 25*(6), suppl.: 26.

DOBROVOLSKI, A. A., 1969. The life-cycle of *Paralepoderma cloacicola* (Lühe, 1909) Dollfus, 1950 (Trematoda : Plagiorchiidae). *Vest. leningr. gos. Univ., Ser. Biol., 24*(9): 28-38 (in Russian).

EFFORD, I. E. & TSUMURA, K., 1969. Observations on the biology of the trematode *Megalodiscus microphagus* in amphibians from Marion lake, British Columbia. *Am. Midl. Nat., 81*(1): 197-203.

ETGES, F. J., 1953. Observations on excystation of metacercariae of *Plagitura salamandra* Holl, 1928, with some notes on its morphology. *J. Parasit., 39*(5): 568-569.

GALLIEN, L., 1935. Recherches expérimentales sur le dimorphisme évolutif et la biologie de *Polystomum integerrimum* Fröhl. *Trav. Stan. zool. Wimereux, 12*(1): 1-181.

GOODCHILD, C. G., 1943. The life-history of *Phyllodistomum solidum* Rankin, 1937, with observations on the morphology, development and taxonomy of Gorgoderinae (Trematoda). *Biol. Bull. mar. biol. Lab. Woods Hole, 84*(1): 59-86

GRABDA, B., 1959. The life-cycle of *Astiotrema trituri* B. Grabda, 1959 (Trematoda, Plagiorchiidae). *Acta parasit. pol., 7*(24): 489-498.

GRABDA-KAZUBSKA, B., 1960. Life-cycle of *Haematoloechus similis* (Looss, 1899) (Trematoda, Plagiorchiidae). *Acta parasit. pol., 8*(21/32): 357-367.

GRABDA-KAZUBSKA, B., 1963. The life-cycle of *Metaleptophallus gracillimus* (Lühe, 1909) and some observations on the biology and morphology of developmental stages of *Leptophallus nigrovenosus* (Bellingham, 1844). *Acta parasit. pol., 11*(25): 349-370.

GRABDA-KAZUBSKA, B., 1969. Studies on abbreviation of the life-cycle in *Opisthioglyphe ranae* (Fröhlich, 1791) and *O. rastellus* (Olsson, 1876) (Trematoda : Plagiorchiidae). *Acta parasit. pol., 16*(27): 249-269.

GRABDA-KAZUBSKA, B., 1970. Studies on the life-cycle of *Haplometra cylindracea (Zeder, 1800)* (Trematoda : Plagiorchiinae). *Acta parasit. pol., 18*(44): 497-512.

HARRIS, A. H., HARKEMA, R. & MILLER, G. C., 1970. Life-cycle of *Procyotrema marsupiformis* Harkema and Miller (Trematoda : Strigeoidea : Diplostomatidae). *J. Parasit., 56*(2): 297-301.

HUGUES, R. C., 1929. Studies on the trematode family Strigeidae (Holostomidae) No. XIV. Two new species of Diplostomula. *Occ. Pap. Mus. Zool. Univ. Mich., 202*: 1-28.

JOHNSTON, T. H. & ANGEL, L. M., 1941. The life-history of *Echinostoma revolutum* in South Australia. *Trans. R. Soc. S. Aust., 65*(2): 317-322.

JORDAN, H. E. & BYRD, E. E., 1967. The life-cycle of *Brachycoelium mesorchium* Byrd, 1937 (Trematoda : Brachycoeliinae). *Z. Parasitenk., 29*(1): 61-84.

KRULL, W. H., 1931. Life-history studies on two frog lung flukes, *Pneumonoeces medioplexus* and *Pneumonoeces parviplexus. Trans. Am. microsc. Soc., 50*(3): 215-277.

KRULL, W. H., 1933. Studies on the life-history of a frog lung fluke *Haematoloechus complexus* (Seely, 1906). *Z. Parasitenk., 6*(2): 192-206.

KRULL, W. H., 1935. Studies on the life-history of a bladder fluke, *Gorgodera amplicava* Looss, 1899. *Pap. Mich. Acad. Sci., 20:* 697-710.

KRULL, W. H. & PRICE, H. F., 1932. Studies on the life-history of *Diplodiscus temperatus* Stafford, from the frog. *Occ. Pap. Mus. Zool. Univ. Mich. 237:* 1-38.

LANG, A., 1892. Über die Cercarie von *Amphistoma subclavatum. Ber naturf. Ges. Freiburg i. B., 6*(3): 81-89.

LANG, A., 1968. The life-cycle of *Cephalogonimus americanus* Stafford, 1902 (Trematoda : Cephalogonimidae). *J. Parasit., 54*(5): 945-949.

LEIGH, W. H., 1946. Experimental studies on the life cycle of *Glypthelmins quieta* (Staff., 1900), a trematode of frogs. *Am. Midl. Nat., 35*(2): 460-483.

MARTIN, G. W., 1969. Description and life cycle of *Glypthelmins hyloreus* sp. n. (Digenea : Plagiorchiidae). *J. Parasit., 55*(4): 747-752.

MATHIAS, P. & VIGNAUD, P., 1935. Sur le cycle évolutif d'un trématode de la sous-famille des Pleurogenetinae Looss (*Pleurogenes claviger* Rud.). *C. r. Séanc. Soc. Biol., 120*(32): 397-398.

MIZELLE, J. D., KRITSKY, D. C. & BURY, R. B., 1968. Studies on monogenetic trematodes XLI. *Gyrodactylus ensatus* sp.n. The first species of the genus described from Amphibia. *J. Parasit., 54:* 281-282.

MIZELLE, J. D., KRITSKY, D. C. & McDOUGAL, H. D., 1969. Studies on monogenetic trematodes XLII. New species of *Gyrodactylus* from Amphibia. *J. Parasit., 55:* 740-741.

NAJARIAN, H. H., 1954. Developmental stages in the life-cycle of *Echinoparyphium flexum* (Linton, 1892) Dietz, 1910. *J. Morph., 94*(1): 165-197.

NEUHAUS, W., 1940. Entwicklung und Biologie von *Pleurogenes medians* Olss. *Zool. Jb. (Syst.), 73*(3): 207-242.

NOBLE, E. R., 1966. Dependence of the parasitic fauna of fish and amphibia upon ecological peculiarities of the host. *Int. Congr. Parasit. (1st), Rome,* 550-551.

OCHI, S., 1930. Uber die Entwicklungsgeschichte von *Mesocoelium brevicaecum* n. sp. *J. Okayama med. Soc.,* 42(2): 388-402 (in Japanese).

ODENING, K., 1965. Der Entwicklungszyklus von *Parastrigea robusta* Szidat, 1928 (Trematoda, Strigeida) in Raum Berlin. *Z. Parasitenk.,* 26: 185-196.

ODENING, K., 1965. Der Lebenszyklus von *Neodiplostomum spathoides* Dubois (Trematoda, Strigeida) in Raum Berlin nebst Beiträgen zur Entwicklungsweise verwandter Arten. *Zool. Jb., Abt. System., Geogr. u. Okol.,* 92: 523-524.

ODENING, K., 1966. The life cycle of *Codonocephalus urniger* (Rudolphi) (Trematoda). *Int. Congr. Parasit. (1st),* Rome: 528-529.

ODENING, K., 1967. Die Lebenszyklen von *Strigea falconispalombi* (Viborg), *S. strigis* (Schrank) und *S. sphaerula* (Rudolphi) in Raum Berlin. *Zool. Jb., Abt. System., Geogr. u. Okol.,* 94: 1-67.

OKABE, K., 1937. On the life history of a frog trematode *Loxogenes liberum* Seno. *Annotnes zool. jap.,* 16(1): 42-52.

OLIVIER, L. J., 1940. Life history studies of two strigeid trematodes of the Douglas Lake region, Michigan, *J. Parasit.,* 26: 447-477.

OLSEN, O. W., 1937. Description and life history of the trematode *Haplometrana utahensis* sp. nov. (Plagiorchiidae), from *Rana pretiosa. J. Parasit.,* 23(1): 13-28.

OZAKI, Y. & OKUDA, Y., 1951. Preliminary note on the life history of *Liolope copulans* Cohn. *J. Sci. Hiroshima Univ. (Ser. B, Div. 1),* 12(10): 113-119.

PEARSON, J. C., 1956. Studies on the life cycle and morphology of the larval stages of *Alaria arisaemoides* Augustine and Uribe, 1927 and *Alaria canis* La Rue and Fallis, 1936 (Trematoda : Diplostomidae). *Can. J. Zool.,* 34: 295-387.

PEARSON, J. C., 1959. Observations on the morphology and life cycle of *Strigea elegans* Chandler and Rausch, 1947 (Trematoda : Stringeidae). *J. Parasit.,* 45(2): 155-174.

PLASOTA, K., 1969. The effect of some ecological factors on the parasitofauna of frogs. *Acta. parasit. pol.,* 16(6): 47-60.

RANKIN, J. S., JR., 1939. Life-cycle of the frog bladder fluke, *Gorgoderina attenuata* Stafford, 1902. *Am. Midl. Nat.,* 21(2): 476-488.

RANKIN, J. S., JR., 1944. A review of the trematode genus *Halipegus* Looss, 1899, with an account of the life history of *H. amherstensis* n. sp. *Trans. Am. microsc. Soc.,* 63(2): 149-164.

SAVINOV, V. A., 1953. Development of *Alaria alata* (Goeze, 1781) in the organism of the definitive and reservoir hosts. *Contributions to Helminthology.* Published to commemorate the 75th birthday of K. I. Skrjabin. 611-616 (in Russian).

SHEVCHENKO, N. N. & VERGUN, G. I., 1961. Life-cycle of *Prosotocus confusus* (Looss, 1894) Looss, 1899 of amphibians. *Helminthologia,* 3(1/4): 294-298 (in Russian).

SHIBUE, H., 1953. The first intermediate host of a frog trematode *Pleurogenes japonicus* Yamaguti. *Jap. J. med. Sci. Biol.,* 6(2): 213-220.

STUNKARD, H. W., 1936. The morphology and life history of *Plagitura parva* Stunkard, 1933 (Trematoda). *J. Parasit.,* 22(4): 354-374.

TALBOT, S. B., 1933. Life history studies of trematodes of the subfamily Reniferinae. *Parasitology,* 25(4): 518-545.

THOMAS, L. J., 1934. Further studies on the life cycle of a frog tapeworm, *Ophiotaenia saphena* Osler. *J. Parasit.,* 20: 291-294.

THOMAS, L. J., 1937. Environmental relations and life history of the tapeworm, *Bothriocephalus rarus* Thomas. *J. Parasit.,* 23: 133-152.

THOMAS, L. J., 1939. Life-cycle of a fluke, *Halipegus eccentricus* n. sp. found in the ears of frogs. *J. Parasit.,* 25(3): 207-221.

THOMAS, L. J. & JOHNSON, A. D., 1934. Experiments and observations on the life cycle of *Halipegus occidualis. J. Parasit.,* 20(6): 327.

THOMAS, J. D., 1965. The anatomy, life history and size allometry of *Mesocoelium monodi* Dollfus, 1929. *J. Zool., Lond.,* 146: 413-446.

THURSTON, J. P., 1964. The morphology and life cycle of *Protopolystoma xenopi* (Price) Bychovsky in Uganda. *Parasitology,* 54: 441-450.

THURSTON, J. P., 1967. The morphology and life cycle of *Cephalochlamys namaquensis* (Cohn, 1906) (Cestoda : Pseudophyllidea) from *Xenopus muelleri* and *X. laevis. Parasitology,* 57: 187-200.

TIMON-DAVID, J., 1966. Nouvelles données sur la biologie et la systématique de *Massaliatrema gyrinicola* R. Ph. Dollfus et J. Timon-David, 1960 (Trematoda, Digenea, Heterophyidae). *Int. Congr. Parasit. (1st), Rome:* 545-546.

VOJTKOVA, L., 1966. Zur Kenntnis des Entwicklungszyklus von *Holostephanus volgensis* (Sudarikov, 1962) n. comb. (Trematoda, Digenea : Cyathocotylidae). *Věst. čsl. Spol. zool.* 30(3): 275-286.

YAMAGUTI, S., 1940. Über der Infektionsmodus von *Diplorchis ranae* Ozaki, 1931. *Z. Parasitenk.,* 12(1): 84-85.

YAMAGUTI, S., 1943. Life history of a frog tapeworm *Ophiotaenia ranae* Yamaguti, 1938. *Jap. J. Zool.,* *10*(3): 455-460.

YOUNG, S., 1934. Ueber das Wachstum der *Diphyllobothrium mansoni* (Cobbold, 1883) Joyeux, 1928, im Darme des Endwirtes (Hundes) und die von diesen Bandwurm hervorgerufene Anämie. *J. Shanghai Sci. Inst., 3*(3): 51-113.

DISCUSSION

J. D. Smyth

Would you comment on the importance of neotenic forms in the infective process in relation to the behaviour of amphibian hosts?

C. Combes

According to the latest research, it seems that neoteny is a general phenomenon with the genus *Polystoma* (Miss A.-M. Maeder of the University of Neufchâtel, has recently discovered neotenic forms in African polystomes). It can be thought, however, that the quantitative importance of the neotenic form in the life-cycle of the parasites is closely related to the length of its life, hence to the length of the larval development of the amphibian. Unfortunately, research in the quantitative field is to this day too scanty to establish a correlation. I am working on this subject.

J. D. Smyth

In one of your papers you mention a polystome whose neotenic larva (in contrast to *P. intergerrimum*) produces a neotenic egg. Is this developmental pattern related to different behaviour patterns in these different hosts?

C. Combes

It seems that with *Polystoma integerrimum* (a parasite of *Rana temporaria*) the larvae born of neotenic forms cannot themselves develop into neotenics. However, we have shown in 1968 that this is possible with *Polystoma pelobatis* (a parasite of *Pelobates cultripes*) and very recently we have had the same positive result with *Polystoma gallieni* (a parasite of *Hyla meridionalis*). We thus think that the possibility for the larva born of a neotenic to grow into a neotenic itself is perhaps a general phenomenon. But this possibility is used under natural conditions only if the tadpoles remain such for a long time or if several successive spawnings occur in a perennial biotope. It seems indeed that these conditions exist rather in the case of *P. pelobatis* or *P. gallieni* but one cannot be positive about this.

R. M. Cable

While the life-cycles of trematodes fall into the categories given here, it should be borne in mind that, as far as a taxon is concerned, a variety of patterns can occur although one predominates. An excellent example is the Brachylaimidae with *Leucochloridium* at one extreme and *Leucochloridiomorpha* with its free living, fork-tailed cercaria and three-host cycle at the other.

C. Combes

I think it is very important indeed to note that a classification of the means of transmission can be entirely independent from a taxonomical classification. One must not be surprised that within the same family or the same genus, different food-chains (whether simple or complex) are used by different species.

LIST OF LIFE-CYCLES WHICH MAY BE RELATED TO THE DIFFERENT MODES OF INFESTATION

The life-cycles mentioned are those for which sufficient data are available; however, those for which a controversy in the mode of infestation of the amphibian exists are not listed.

Each cycle is related to the mode of infestation which *seems* to be the main one for it.

The work mentioned for each platyhelminth is either the most complete, or the most recent, when this is of any particular interest.

ABBREVIATIONS

A	Ancylidae	Jap.	Japan	Pl	Planorbidae
Afr.	Africa	La	Lamellibranchiata	Pr	Prosobranchia
Amph.	Amphibian	Li	Lynnaeidae	Rept.	Reptile
Cosm.	cosmopolitan	NA	North America	S	Succineidae
E	Eulotidae	P	Pulmonata	t	tadpole
Eur.	Europa	Ph	Physidae	unp.	unpublished
G	Gastropoda			Z	Zonitidae

Type A1

Polystoma integerrimum (Fröhlich, 1798)			; *Rana temporaria*	; Eur.	; Gallien, 1935
Polystoma gallieni Price, 1938			; *Hyla meridionalis*	; Eur.	; Combes, 1968
Polystoma pelobatis Euzet & Combes, 1965			; *Pelobates cultripes*	; Eur.	; Combes, 1968
Gyrodactylus ensatus Mizelle, Kritshy & Bury, 1968			; *Dicamptodon ensatus*	; NA	; Mizelle *et al.*, 1968
Gyrodactylus ambystomae Mizelle, Kritshy & McDougal, 1969			; *Ambystoma macrodactylum*	; NA	; Mizelle *et al.*, 1969
Gyrodactylus aurorae Mizelle, Kritshy & McDougal, 1969			; *Rana aurora*	; NA	; Mizelle *et al.*, 1969

Type A2

Opisthioglyphe rastellus (Olsson, 1876)	; G P Li	; /	; *Rana temporaria*, etc.	; Eur.	; Grabda-Kazubska, 1969
Haplometra cylindracea (Zeder, 1800)	; G P Li	; /	; *Rana temporaria*, etc.	; Eur.	; Grabda-Kazubska, 1970
Glypthelmyns hyloreus Martin, 1969	; G P Li	; /	; *Hyla regila*	; NA	; Martin, 1969

Type A3

Procyotrema marsupiformis Harkema & Miller, 1959	; G P Pl	; Amph.	; Mammal	; NA	; Harris *et al.*, 1970
Tylodelphys excavata (Rudolphi, 1803)	; G P Pl	; Amph.	; Bird	; Eur.	; Hugues, 1929
Diplostomum micradenum Olivier, 1940	; G P Li	; Amph.	; Bird	; NA	; Olivier, 1940
Parastrigea robusta Szidat, 1928	; G P Pl	; Amph.	; Bird	; Eur.	; Odening, 1965
Neodeplodistomum spatboides Dubois, 1937	; G P Pl	; Amph.	; Bird	; Eur.	; Odening, 1965

Type A4

Taxon	Code	Intermediate	Final host	Distribution	Reference
Strigea elegans Chandler, 1947	; G P Pl	; Amph. then Rept. or Birds	; Bird	; NA	; Pearson, 1959
Alaria arisaemoides Augustine & Uribe, 1927	; G P Pl	; Amph. then Rept.	; Mammal	; NA	; Pearson, 1956
Alaria canis La Rue & Fallis, 1936	; G P Pl	; Amph. then Rept.	; Mammal	; NA	; Pearson, 1956
Alaria alata (Goeze, 1782)	; G P Pl	; Amph. then Rept. or Mammal	; Mammal	; Eur.	; Savinov, 1953
Alaria mustelae Bosma, 1931	; G P Pl	; Amph. then Rept. or Mammal	; Mammal	; NA	; Bosma, 1934

Type A5

Taxon	Code	Intermediate	Final host	Distribution	Reference
Leptophallus nigrovenosus (Bellinghan, 1844)	; G P Li	; Amph. (t)	; Reptile	; Eur.	; Grabda-Kazubska, 1963
Metaleptophallus gracillimus (Lühe, 1909)	; G P Li	; Amph. (t)	; Reptile	; Eur.	; Grabda-Kazubska, 1963
Paralepoderma cloacicola (Lühe, 1909)	; G P Pl	; Amph. (t)	; Reptile	; Eur.	; Dobrovolski, 1969
Lecithiorchis primus (Stafford, 1905)	; G P Ph	; Amph. (t)	; Reptile	; NA	; Talbot, 1933
Holostephanus volgensis (Sudarikov, 1962)	; G Pr	; Amph. (t)	; Bird	; Eur.	; Vojtkova, 1966
Massialiatrema gyrinicola Dollfus & Timon-David, 1960	; G Pr	; Amph. (t)	; Bird	; Eur.	; Timon-David, 1966
Echinostoma revolutum (Fröhlich, 1802)	; G P Pl or Ph or Li	; Amph. (t) (+Molluscs)	; Bird	; Cosm.	; Johnston & Angel, 1941
Echinoparyphium ellisi (Johnston & Simpson, 1944)	; G P Li	; Amph. (t)	; Bird	; Aus.	; Johnston & Angel, 1949
Echinoparyphium flexum (Linton, 1892)	; G P Ph	; Amph. (t)	; Bird	; NA	; Najarian, 1954
Echinoparyphium recurvatum (Linstow, 1873)	; G P Pl or Ph or Li	; Amph. (t) (+Molluscs)	; Bird or Mammal	; Cosm.	; Bittner, 1925

Type B1

Taxon	Code	Intermediate	Final host	Distribution	Reference
Sphyranura oligorchis Alvey, 1933			; *Necturus maculosus*	; NA	; Alvey, 1936

Type B2

Taxon	Code	Intermediate	Final host	Distribution	Reference
Euzetrema knoepffleri Combes, 1965			; *Euproctus montanus*	; Eur.	; Combes et al. (unp.)
Protopolystoma xenopi (Price, 1943)			; *Xenopus laevis*, etc.	; Afr.	; Thurston, 1964
Diplorchis ranae Ozaki, 1931			; *Rana rugosa*	; Jap.	; Yamaguti, 1940
Gyrdicotylus gallieni Vercammen-Grandjean, 1960			; *Xenopus laevis*	; Afr.	; Tinsley (in press)

Type B3

Taxon	Code	Intermediate	Final host	Distribution	Reference
Opisthioglyphe ranae (Fröhlich, 1971)	; G P Li	; /	; *Rana esculenta*, etc.	; Eur.	; Grabda-Kazubska, 1969

Type B4

Glyptbelmyns quieta (Stafford, 1900)	; *Rana catesbiana*, etc.	; G P Ph	; /	; NA	; Leigh, 1946
Haplometrana utabensis Olsen, 1937	; *Rana pretiosa*, etc.	; G P Ph	; /	; NA	; Olsen, 1937
Megalodiscus temperatus (Stafford, 1905)	; *Rana, Bufo*, etc.	; G P Pl or A	; /	; NA	; Krull & Price, 1932
Megalodiscus micropbagus Ingles, 1936	; *Ambystoma gracile*, etc.	; G P Pl	; /	; NA	; Efford & Tsumura, 1969
Diplodiscus subclavatus (Pallas, 1760)	; *Rana :sculenta*, etc.	; G P Pl	; /	; Eur.	; Lang, 1892

Type B5

Eurybelmys squamula (Rudolphi, 1819)	; Mammal	; G Pr	; Amph.	;	; Anderson & Pratt, 1965
Ratzia parva (Stossich, 1904)	; Reptile	; G Pr	; Amph.	;	; Buttner, 1952

Type C1

Halipegus eccentricus Thomas, 1939	; *Rana catesbiana*, etc.	; G P Pl or Ph	; Copepoda	; NA	; Thomas, 1939
Halipegus occidualis Stafford, 1905	; *Rana clamitans*, etc.	; G P Pl or Ph	; Copepoda	; NA	; Thomas & Johnson, 1934
Halipegus amberstensis Laballero, 1947	; *Rana clamitans*, etc.	; G P Ph	; Copepoda	; NA	; Rankin, 1944
Halipegus ovocaudatus (Vulpian, 1958)	; *Rana esculenta*, etc.	; G P Pl	; Insect larva	; Eur.	; Sinitzin, 1905

Type C2

Opbiotaenia sapbena Osler, 1931	; *Rana clamitans*	; Copepoda;		; NA	; Thomas, 1934
Opbiotaenia ranae Yamaguti, 1938	; *Rana nigromaculata*	; Copepoda;		; Jap.	; Yamaguti, 1943

Type C3

Spirometra erinacei-europaei (Rudolphi, 1819)	; Mammal	; Copepoda; Amph.		; Cosm.	; Young, 1934

Type C4

Astiotrema trituri Grabda, 1959	; *Triturus vulgaris*	; G P Pl	; Cladocera	; Eur.	; Grabda, 1959
Liolope copulans Cohn, 1902	; *Megalobatrachus japonicus*	; unknown	; Fish	; Jap.	; Ozaki & Okuda, 1951
Manodistomum parvum (Stunkard, 1933)	; *Triturus viridescens*	; G P Pl	; Mollusc	; NA	; Stunkard, 1936
Manodistomum salamandra (Holl, 1928)	; *Triturus viridescens*	; G P S	; Mollusc	; NA	; Etges, 1953
Pleurogenoides japonicus (Yamaguti, 1936)	; *Rana nigromaculata*	; G P Ph	; Decapoda	; Jap.	; Shibue, 1953

Type C5

Cepbalocblamys numaquensis (Cohn, 1906)	; *Xenopus muelleri*, etc.	; Copepoda;		; Afr.	; Thurston, 1967
Botbriocepbalus rarus Thomas, 1937	; *Triturus viridescens*	; Copepoda;		; NA	; Thomas, 1937

Type D1

Haematoloechus similis Looss, 1899	; G P Pl	; Insect	; *Rana esculenta*, etc.	; Eur.	; Grabda-Kazubska, 1960
Haematoloechus medioplexus Stafford, 1902	; G P Pl	; Insect	; *Rana pipiens*, etc.	; NA	; Krull, 1931
Haematoloechus complexus (Seely, 1906)	; G P S	; Insect	; *Rana pipiens*, etc.	; NA	; Krull, 1933
Haematoloechus parviplexus (Irwin, 1929)	; G P Pl	; Insect	; *Rana clamitans*	; NA	; Krull, 1931
Haematoloechus pyrenaicus Combes, 1965	; G P A	; Insect	; *Rana temporaria*, etc.	; Eur.	; Combes, 1968
Pleurogenes claviger (Rudolphi, 1819)	; G Pr	; Insect	; *Rana esculenta*, etc.	; Eur.	; Mathias & Vignaud, 1935
Pleurogenoides medians (Olsson, 1876)	; G Pr	; Insect	; *Rana esculenta*, etc.	; Eur.	; Neuhaus, 1940
Prosotocus confusus (Looss, 1894)	; G Pr	; Insect	; *Rana esculenta*, etc.	; Eur.	; Shevchenko & Vergun, 1961
Loxogenes liberium Seno, 1908	; G P Ph	; Insect	; *Rana nigromaculata*, etc.	; Jap.	; Okabe, 1937
Plagiorchis ramlianus (Looss, 1896)	; G P Ph	; Insect	; *Rana mascareniensis*	; Afr.	; Abdel Azim, 1935
Cephalouterina dicamptodoni Senger & Macy, 1953	; G Pr	; Insect	; *Dicamptodon, Ascaphus*	; NA	; Anderson *et al.*, 1966
Bunoderella metteri Schell, 1964	; L	; Insect	; *Ascaphus truei*	; NA	; Anderson *et al.*, 1965
Gorgodera euzeti Lees & Combes, 1968	; L	; Insect	; *Rana temporaria*, etc.	; Eur.	; Combes, 1968
Phyllodistomum solidum Rankin, 1937	; L	; Insect	; *Desmognathus fusca*	; NA	; Goodchild, 1943
Phyllodistomum americanum Osborn, 1903	; L	; Insect	; *Ambystoma tigrinum*	; NA	; Crawford, 1939
Tetracheilos ascaphi Pratt, 1964	; L	; Insect	; *Ascaphus truei*	; NA	; Anderson, 1964

Type D2

Cephalogonimus americanus Stafford, 1902	; G P Pl	; Tadpoles	; *Rana clamitans*	; NA	; Lang, 1968
Cephalogonimus europaeus Blaizot, 1910	; G P Li	; Tadpoles	; *Rana ridibunda*, etc.	; Eur.	; Combes *et al.* (unp.)

Type D3

Gorgoderina vitelliloba (Olsson, 1876)	; L	; Tadpoles	; *Rana temporaria*, etc.	; Eur.	; Combes, 1968
Gorgoderina attenuata (Stafford, 1902)	; L	; Tadpoles	; *Rana catesbiana*, etc.	; NA	; Rankin, 1939
Gorgodera amplicava Looss, 1899	; L	; Tadpoles	; *Rana, Bufo*, etc.	; NA	; Krull, 1935

Type D4

Codonocephalus urniger (Rudolphi, 1819)	; G P Li	; Amph.	; Bird	; Eur.	; Odening, 1966
Strigea sphaerula (Rudolphi, 1803)	; G P Pl	; Amph.	; Bird	; Eur.	; Odening, 1967
Strigea strigis (Schrank, 1788)	; G P Pl	; Amph.	; Bird	; Eur.	; Odening, 1967

Type E1

Brachycoelium mesorchium Byrd, 1937	; G P M	; Terrestrial Molluscs	; *Eurycea, Desmognathus*	; NA	; Jordan & Byrd, 1967

Type E2

Brachycoelium obesum Nicoll, 1914	; G P Z	; /	; *Plethodon glutinosus*, etc.	; NA	; Cheng, 1960
Mesocoelium brevicaecum Ochi, 1930	; G P E	; /	; *Bufo, Rana*, etc.	; Jap.	; Ochi, 1930
Mesocoelium monodi Dollfus, 1929	; G P E	; /	; *Bufo, Rana*, etc.	; Afr.	; Thomas, 1965

The sense organs of trematode miracidia

B. E. BROOKER

The Nuffield Institute of Comparative Medicine, The Zoological Society of London, London

Simple ciliated nerve endings, believed to be tangoreceptors, are arranged in two sets around the apical papilla of *Schistosoma mansoni* miracidia and a single set of ciliated pit nerve endings, suggested chemoreceptors, is found just posterior to them. The lateral papillae in the miracidium of this species and of *Diplostomum spathaceum* contain a single bulbous nerve ending and lie immediately posterior to a ciliary receptor. Mechanical interaction between these two structures may provide the larva with information about its orientation with respect to gravity. In the larva of *D. spathaceum* two dorsal ciliated papillae are believed to be chemoreceptors. Structures quite distinct from eyespots are found in the miracidium of both species. Each one of these cells contains an internally ciliated vacuole occupied to a large extent by lamellar stacks of ciliary membrane.

CONTENTS

INTRODUCTION

The chief biological function of the trematode miracidium is to enter a suitable molluscan host where it may eventually give rise to succeeding generations of larvae. In some cases where infection of the mollusc is achieved passively, the miracidium is ingested whilst still in the egg and only emerges when in the gut of the host but in many species a free-living motile larva actively locates and penetrates the snail and it is with the sense organs of these miracidia that this paper is primarily concerned. A number of authors have reported photo- and geotactic responses by miracidia and although the importance of chemotaxis in host location is recognized, whether it is in any way related to host specificity is still being argued. In many cases it appears that the initial response of the larva to environmental stimuli is such that it moves not to a particular host but to a certain type of environment in which potential hosts are more likely to be found than in any other. This process, which Salt (1935) referred to as the ecological selection of hosts, is probably succeeded by a more specific search for suitable hosts in which chemotaxis may play a prominent part. Thus the miracidia of *Schistosoma japonicum*, which are positively phototactic and negatively geotactic, tend to swim to the surface of the water (Faust, 1924) where they are more likely to encounter their amphibious intermediate hosts, *Oncomelania* spp., which are frequently found near the margins of the water. On the other hand, the experiments performed by Shiff (1969) in which miracidia of *S. haematobium* were exposed to cages of snails tethered at different depths in a pond, one half of which was in the shade and the other in bright sunlight, suggest that these larvae are negatively phototactic and positively geotactic. The advantage of such behaviour is apparent when it is considered that in at least some situations the probable molluscan hosts, *Bulinus* spp., are bottom feeders (Wright, 1962). However, to attribute any set pattern of behavioural responses to the larva of a particular species is probably an oversimplification. Takahashi, Mori & Shigeta (1961) have shown that the phototactic and to less extent the geotactic responses of *S. japonicum* miracidia can be modified by exposure to certain combinations of light intensity and water temperature. It is tempting to suggest that such changes correspond under natural conditions to periodic fluctuations in the vertical distribution of the snail hosts. An example of how the innate phototactic behaviour may change with time is given by Isseroff & Cable (1968) for the miracidium of *Philophthalmus megalurus.* Although the snail host for this larva is a bottom dweller, the miracidia are strongly photopositive immediately after hatching and swim to the surface of the water. With age, their behaviour changes and they begin to swim downwards. The authors suggest that the initial period of photosensitive behaviour, during which time they make few attempts to penetrate a snail, corresponds under natural conditions to a period of dispersal after which they move to the bottom to find a suitable host.

Although structures believed to be sensory receptors have been reported from several trematodes and trematode larvae, in no case is there experimental evidence to support their postulated function. This is largely because they are usually so small and difficult to locate by light microscopy that normal electro-physiological and ablation techniques are not easy to use. Identification of function is usually based on morphological similarities to receptors of other invertebrates where the function is either known or suspected. However it should be remembered that this practice is fraught with difficulties and allows only tentative conclusions to be drawn.

TEGUMENTARY SENSE ORGANS

Because most of the sense organs of trematode miracidia occur near or at the surface of the larva it is necessary, before considering their structure, to mention briefly some aspects of the organization of the larval tegument. The greater part of the body is covered by several tiers of ciliated plates and each plate is separated from its neighbour by a narrow aciliated area. In most miracidia it appears that this aciliated area or intercellular ridge is the superficial portion of a cell, the major part of which lies below the general surface of the larva. In some species, for example *Schistosoma mansoni*, the ciliated plates are similarly only superficial extensions of a deeper cell so that by light microscopy they appear to be anucleate. However in other species, as Wilson (1969) has shown in the miracidium of *Fasciola hepatica,* the ciliated plates are discrete nucleated cells. With the possible exception of the apical papilla, where receptors reach the surface they usually do so by penetrating the intercellular ridge and are secured to it by septate desmosomes.

The greatest concentration of receptors in those miracidia which have been examined is found on the apical papilla. In the miracidium of *Schistosoma mansoni* two types of ciliated nerve endings are arranged around the apical papilla in three sets. Two anterior sets are each composed of four to six receptors each of which bears a single terminal cilium (Fig. 1; Plate 1A, B) and the third set, which lies immediately anterior to the lateral gland cell ducts, contains about six ciliated pit nerve endings (Fig. 1; Plates 1A and 2A). It is interesting that each and every one of the nerve endings and lateral gland cell ducts is embedded in a considerable posterior extension of the apical gland cell (Fig. 1; Plates 1A and 2A). However, it appears that the apical papillae of miracidia do not always possess both receptor types, for Wilson (1970) was able to find only ciliated pits in the case of *Fasciola hepatica*. In her work on monogenean oncomiracidia, Lyons (1969) has argued that ciliated pit nerve endings are chemosensory. Their position on the apical papilla of the miracidium and at the anterior tip of the cercaria of *S. mansoni* (Morris, 1971),

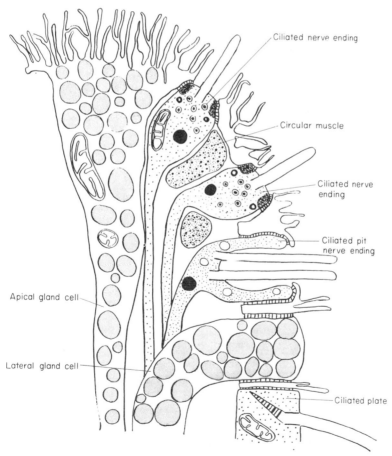

Figure 1. Diagrammatic reconstruction of one half of the apical papilla of the miracidium of *Schistosoma mansoni*

the points at which initial contact with the hosts are made, certainly suggests
that they are responsible for either chemo- or tangoreception. However, it may
be argued that the presence of ciliated pits in an olfactory organ of
cephalopods (the rhinophore) (Barber & Wright, 1969) where they are believed
to be chemosensory, points to a similar function in trematode larvae. It may be
that whereas some of the ciliated pits are involved in host location, others are
concerned with some kind of host identification. Nerve endings bearing a single
terminal cilium have been described from several larval and adult trematodes
and although minor differences have been recorded in different species, they
are all basically similar in appearance (Dixon & Mercer, 1965; Erasmus, 1967;
Matricon-Gondran, 1971). Most authors believe that they are tangoreceptors.
From their anterior position it seems probable that they fulfil a similar
function in the miracidium of *S. mansoni,* but just how much information they
provide about the surface with which they come into contact is unknown. It
would appear therefore that in *S. mansoni* miracidia host location and
presumably the decision to penetrate involves both tactile and chemo-stimula-
tion. It may be that in *F. hepatica* where obvious tangoreceptors have not yet
been found host location and penetration involves only chemoreception. The
ability, in some fasciolids, to induce ciliated plate shedding, a process which
normally accompanies penetration, by the addition of snail extract may be of
relevance here.

    In addition to these receptors on the apical papilla, the miracidia of
Schistosoma mansoni possess two rings of ciliated nerve endings between the
second and third series of ciliated plates. The bulbous tip of each ending bears
several very long cilia which are often observed to contain two or more
axonemes (Plate 1C). Their function is unknown.

    The miracidia of most trematodes possess a pair of conspicuous lateral
papillae located between the first and second series of ciliated plates. A
connective between each papilla and the nerve mass was identified by some
earlier workers as a glandular duct carrying secretions to the outside because in
many of the species examined substances were seen to be extruded from the
papillae (Cort, 1919; Stunkard, 1923; Faust & Meleney, 1924; Wall, 1941;
Ottolina, 1957). Other authors identified this connective as a nerve and
therefore considered the papillae to be sensory (Reisinger, 1923; Lynch, 1934;
Bennett, 1936). Rees (1940) working with the miracidium of *Parorchis acanthus*
reported that the lateral papillae were capable of considerable extension and
although suggesting that they played a part in the emergence of the miracidium
from the egg, she also believed them to be sensory. More recently, the fine
structure of the papillae in the miracidium of *Fasciola hepatica* has been
described by Wilson (1970) who found that each one consisted of two distinct
nerve fibres protruding above the general surface of the larva and covered by an
evagination of the intercellular ridge. In the miracidia of *Diplostomum
spathaceum* and *Schistosoma mansoni,* the papillae are essentially similar
except that each is composed of only one bulbous nerve ending (Plates 2B and
3A). A striking feature of these endings is that they contain neuronal
microtubules interspersed between numerous small vesicles. It is assumed that
the vesicles are produced in the nerve cell body and then transported to the
nerve ending, a process in which some authors (Smith, Järlfors & Beráner,
1970) have implicated the neuronal microtubules. Because the nerve endings

are separated from the external environment and possess no obvious structure capable of transduction, Wilson (1970) has suggested that the papillae are not sensory. Instead, he proposed that they either represent the developmental phase of a sensory structure or that they are secretory, releasing their enclosed vesicles during penetration of the snail. However, another possibility does exist. A number of ciliated nerve endings are distributed around the body at regular intervals between the first and second tiers of ciliated plates and one of these is always found immediately anterior to each lateral papilla. The structure of these endings appears to be variable, for whereas in *D. spathaceum* they bear one or two cilia containing conspicuous osmiophilic material (Plate 2B), in *F. hepatica* and *S. mansoni* the cilium is surrounded by elaborate membrane evaginations derived from the same cell as that covering the lateral papilla (Plate 3A, B). the close proximity of these cilia to the lateral processes may indicate that the two structures are functionally interrelated. Assuming that the papillae are to some extent flexible and that the ciliated nerve endings are sensory, it can be visualized that mechanical stimulation of the cilia by the papillae will provide the larva with information abouts its orientation with respect to gravity. Wilson's suggestion that in *F. hepatica* the membrane collar surrounding the cilium protects it from low levels of mechanical stimulation such as water currents is compatible with this mechanism. Depending on the specific gravity of their contents, the vesicles of the lateral papillae may be seen as a device for conferring either buoyancy or increased density on the papilla and that whichever of these two possibilities actually exists will depend on the motor effect of stimulating the ciliated endings. Structures almost identical to the lateral papillae have been found on the body and tail of *Cercaria pectinata* (larva of *Bacciger bacciger*) but here they are not associated with a ciliated nerve ending and are considered to be tangoreceptors (Matricon-Gondran, 1971).

In *Diplostomum spathaceum* two dorsal papillae are found between the dorsal ciliated plates of the first tier. In some respects they resemble the lateral papillae, for each one consists of a nerve ending which penetrates an intercellular ridge and protrudes above the general surface of the larva. They contain numerous vesicles but unlike the lateral papillae, they bear a number of cilia and are in direct contact with the external environment (Plates 3C and 4A). The exact number of cilia has not yet been determined but they all project from the nerve ending radially and are orientated parallel to the surface of the miracidium. Each one is about 2-3 μm long, exhibit a "9 + 0" arrangement of microtubules and, like the ciliated nerve endings associated with the lateral papillae, contains droplets of osmiophilic material. In almost every respect the papillae closely resemble the olfactory receptors of vertebrate nasal epithelium (Reese, 1965; De Lorenzo, 1970). Since olfaction is just one facet of chemoreception, it is suggested that the ciliated papillae of *D. spathaceum* miracidia are chemoreceptors. Unlike these papillae, most vertebrate olfactory receptors are covered by a thick layer of mucus in which, it is believed, inhaled odorous substances are dissolved prior to receptor stimulation. In the miracidium, this layer would appear to be unnecessary because the ciliated papillae are in direct contact with the surrounding water. If the suppositions made so far are true, it may be concluded that in some miracidia at least two morphologically distinct types of chemoreceptor are found namely, the ciliated pit nerve endings and the ciliated papillae. Possibly, they correspond to the long and short range receptors postulated by MacInnis (1965).

PHOTORECEPTORS

The eyespots possessed by many miracidia appear to be the only receptors to which a function can be ascribed with any confidence. Fine structural aspects of these supposed photoreceptors have been described by Kümmel (1960), Isseroff (1964) and Isseroff & Cable (1968). In all the species which have been examined, each eyespot consists of a prominent pigment cell possessing one or two concavities occupied by rhabdomeres. In *Fasciola hepatica, Allocreadium lobatum* and a species of *Spirorchis,* each pigment cell contains a pair of rhabdomeres and a fifth receptor occupies a posteromedian extension of the left pigment cell. Although in *Heronimus chelydrae* the situation is slightly complicated by the fact that the left eyespot is composed of two pigment cells, Isseroff & Cable (1968) found that in this species also a total of five rhabdomeres are present. *Philophthalmus megalurus* appears to be exceptional in this respect since only four symmetrically arranged rhabdomeres have been demonstrated (Isseroff, 1964). Although species differences such as the length and orientation of the rhabdomeric microvilli were noted, Isseroff & Cable (1968) suggest that they are related to variations in the phototropic responses which enable the larvae to reach the appropriate host. In *Diplostomum spathaceum,* the organization of the eyespots is very similar to that described from *A. lobatum.* Each pigment cell contains many pigment granules and its concavity is occupied by the parallel microvilli of two rhabdomeres (Plate 4B). Each retinular cell is easily identified because its numerous mitochondria are closely apposed to form one solid mass (Plates 4B and 5). A fifth rhabdomere is located posterior to the two pigment cells but unlike the four rhabdomeres which they contain, its mitochondria are dispersed throughout the cell and the numerous randomly orientated microvilli are surrounded by a thin cytoplasmic extension of the left pigment cell (Plate 6A). This difference in the orientation of the microvilli between the fifth rhabdomere and those occupying the pigment cups has been previously reported from *F. hepatica* and *H. chelydrae* (Isseroff & Cable, 1968). Unless there is some special functional significance to the random orientation of the microvilli, we may be witnessing a stage in the progressive loss of this cell and with it, movement towards the philophthalmid condition.

In addition to the eyespots there is evidence that some miracidia possess structures which, on morphological grounds, appear to be photoreceptors. In the miracidium of *Diplostomum spathaceum,* the retinular cell dendrites emerge from each pigment cup and make contact with a cell which contains a single large vacuole (Plate 5). Embedded in the walls of this vacuole are 10-20 basal bodies which give rise to a corresponding number of cilia projecting into the lumen. The axoneme of each cilium is very short and is composed of a "9 + 0" arrangement of microtubules (Plate 6B). Each cilium possesses an undulatory membrane evagination and the cilia are so arranged that these membrane sheets lie one on top of the other to form lamellar stacks (Plate,7B). Ciliated cells whose cilia possess conspicuous membrane evaginations have been reported from a number of invertebrates (Eakin & Westfall, 1962; Eakin, 1963; Horridge, 1964) and have usually been identified as photoreceptors. Although there is little experimental evidence to support such identification, it is generally supposed that, as in the retinal cells of vertebrates, the folded ciliary membranes are the bearers of photopigment. By analogy with these other invertebrate systems it may be supposed that the two laterally situated ciliated cells in this miracidium are similarly photoreceptive. Wilson (1970) has described two similar structures from the miracidium of *Fasciola hepatica* but in this case they are situated near the lateral papilla and the ciliary membranes are comparatively poorly developed. Wilson suggests that they are gravity receptors but it seems more likely that they are photoreceptors akin to those found in *D. spathaceum.* It is to be expected that, as in the eyespots, some differences in position and the degree to which they are developed will be found. In the miracidium of *Schistosoma mansoni* eyespots are absent but the presence of a photoreceptor has been inferred because of its positively phototropic response to light. It is interesting therefore that internally ciliated cells identical in appearance to those described above are situated below the body wall near the base of the lateral processes (Plate 7A). If these cells are truly photoreceptive, it may well be that they are also to be found in the miracidia of those species which, although lacking eyespots, are photosensitive. In view of their proposed photosensitivity, it is interesting that the only location in which internally ciliated cells resembling those of miracidia are reported to occur is the base of the pineal gland of some mammals (Papacharalampous, Schwink & Wetzstein, 1968).

These observations raise a number of interesting questions. What, for example, is the relative importance of each receptor in those larvae which possess both? Because the photoreceptors of eyespots show an antero-posterior organization, Isseroff & Cable (1968) have suggested that they are concerned with orientation of the larva with respect to light, but the function of the cilium derived receptors seems less obvious especially when it is remembered that the miracidium of *Schistosoma mansoni* appears capable of similar responses to light as those miracidia which possess eyespots. Because of their large surface area of ciliary membrane, the cilium derived receptors may have a greater sensitivity to low levels of illumination than the eyespots and may assume greatest importance at light levels below the threshold of stimulation of the retinular cells.

SUMMARY

Simple ciliated nerve endings, believed to be tangoreceptors, are arranged in two sets around the apical papilla of *Schistosoma mansoni* miracidia and a single set of ciliated pit nerve endings, suggested chemoreceptors, is found just posterior to them. The lateral papillae in the miracidium of this species and of *Diplostomum spathaceum* contain a single bulbous nerve ending and lie immediately posterior to a ciliary receptor. Mechanical interaction between these two structures may provide the larva with information about its orientation with respect to gravity. In the larva of *D. spathaceum,* two dorsal ciliated papillae are believed to be chemoreceptors. Structures quite distinct from eyespots are found in the miracidium of both species. Each one of these cells contains an internally ciliated vacuole occupied to a large extent by lamellar stacks of ciliary membrane.

ACKNOWLEDGEMENTS

I wish to thank Dr C. A. Wright for his constant advice and interest in this project. Part of this work was performed during the tenure of a Senior Research Fellowship at the British Museum (Natural History).

REFERENCES

BARBER, V. C. & WRIGHT, D. E., 1969. The fine structure of the sense organs of the cephalopod mollusc *Nautilus. Z. Zellforsch. mikrosk. Anat., 102:* 293-312.

BENNETT, H. J., 1936. The life history of *Cotylophoron cotylophoron,* a trematode from ruminants. *Illinois biol. Monogr., 14:* 1-119.

CORT, W. W., 1919. Notes on the eggs and miracidia of the human schistosomes. *Univ. Calif. Publs Zool., 18:* 509-519.

DE LORENZO, A. J. D., 1970. In *Taste and Smell in Vertebrates.* Ciba Foundation Symposium. London: Churchill.

DIXON, K. E. & MERCER, E. H., 1965. The fine structure of the nervous system of the cercaria of the liver fluke, *Fasciola hepatica* L. *J. Parasit., 51:* 967-976.

EAKIN, R. M., 1963. In *General physiology of cell specialization.* San Francisco: McGraw-Hill.

EAKIN, R. M. & WESTFALL, J. A., 1962. Fine structure of photoreceptors in the hydromedusan, *Polyorchis penicillatus. Proc. natn. Acad. Sci. U.S.A., 48:* 826-833.

ERASMUS, D. A., 1967. The host-parasite interface of *Cyathocotyle bushiensis* Khan, 1962 (Trematoda : Strigeoidea). II. Electron microscope studies of the tegument. *J. Parasit., 53:* 703-714.

FAUST, E. C., 1924. The reactions of the miracidia of *Schistosoma japonicum* and *S. haematobium* in the presence of their intermediate hosts. *J. Parasit., 10:* 199-204.

FAUST, E. C. & MELENEY, H. E., 1924. Studies on Schistosomiasis Japonica. *Am. J. Hyg. (Monographic Series), 3:* 1-325.

HORRIDGE, G. A., 1964. Presumed photoreceptive cilia in a ctenophore. *Q. Jl microsc. Sci., 105:* 311-317.

ISSEROFF, H., 1964. Fine structure of the eyespot in the miracidium of *Philophthalmus megalurus* (Cort, 1914). *J. Parasit., 50:* 549-554.

ISSEROFF, H. & CABLE, R. M., 1968. Fine structure of photoreceptors in larval trematodes. A comparative study. *Z. Zellforsch. mikrosk. Anat., 86:* 511-534.

KÜMMEL, G., 1960. Die Feinstruktur des Pigmentbecherocells bei Miracidien von *Fasciola hepatica* L. *Zool. Beitr., 5:* 345-354.

LYNCH, J. E., 1934. The miracidium of *Heronimus chelydrae* MacCallum. *Q. Jl microsc. Sci., 76:* 13-33.

LYONS, K. M., 1969. Compound sensilla in monogenean skin parasites. *Parasitology, 59:* 625-636.

MacINNIS, A. J., 1965. Responses of *Schistosoma mansoni* miracidia to chemical attractants. *J. Parasit., 51:* 731-746.

MATRICON-GONDRAN, M., 1971. Etude ultrastructurale des récepteurs sensoriels tégumentaires de quelques Trématodes Digénétiques larvaires. *Z. Parasitenk., 35:* 318-333.

MORRIS, G. P., 1971. The fine structure of the tegument and associated structures of the cercaria of *Schistosoma mansoni. Z. Parasitenk., 36:* 15-31.

OTTOLINA, C., 1957. El miracidio del *Schistosoma mansoni. Revta Sanid. Asist. soc., Caracas, 22:* 1-435.

PAPACHARALAMPOUS, N. X., SCHWINK, A. & WETZSTEIN, R., 1968. Elektronenmikroskopische Untersuchungen am Subcommissuralorgan des Meerschweinchens. *Z. Zellforsch. mikrosk. Anat., 90:* 202-229.

REES, G., 1940. Studies on the germ cell cycle of the digenetic trematode *Parorchis acanthus* Nicoll. Part II. Structure of the miracidium and germinal development in the larval stages. *Parasitology, 32:* 372-391.

REESE, T. S., 1965. Olfactory cilia in the frog. *J. Cell Biol., 25:* 209-230.

REISINGER, E., 1923. Untersuchungen über Bau und Funktion des Excretionsapparates digenetischer Trematoden. *Zool. Anz., 57:* 1-20.

SALT, G., 1935. Experimental studies in insect parasitism. III. Host selection. *Proc. R. Soc. (B), 117:* 413-435.

SHIFF, C. J., 1969. Influence of light and depth on location of *Bulinus (Physopsis) globosus* by miracidia of *Schistosoma haematobium. J. Parasit., 55:* 108-110.

SMITH, D. S., JARLFORS, U. & BERANER, R., 1970. The organisation of synaptic axoplasm in the lamprey (*Petromyzon marinus*) central nervous system. *J. Cell Biol., 46:* 199-219.

STUNKARD, H. W., 1923. Studies on North American blood flukes. *Bull. Am. Mus. nat. Hist., 48:* 165-221.

TAKAHASHI, T., MORI, K. & SHIGETA, Y., 1961. Phototactic, thermotactic and geotactic responses of miracidia of *Schistosoma japonicum. Jap. J. Parasit., 10:* 686-691.

WALL, L. D., 1941. *Spirorchis parvus* (Stunkard). Its life history and the development of its excretory system (Trematoda : Spirorchidae). *Trans. Am. microsc. Soc., 60:* 221-260.

WILSON, R. A., 1969. Fine structure of the tegument of the miracidium of *Fasciola hepatica* L. *J. Parasit., 55:* 124-133.

WILSON, R. A., 1970. Fine structure of the nervous system and specialized nerve endings in the miracidium of *Fasciola hepatica. Parasitology, 60:* 399-410.

WRIGHT, C. A., 1962. In *Bilharziasis.* Ciba Foundation Symposium. London: Churchill.

DISCUSSION

D. Davenport

We can find out a great deal about the sensory control of movement in miracidia even without neuro-physiological techniques. It is quite possible to observe the responses of individual miracidia in detail when they are put under uni- or multi-directional photic stimulation. We should be able to quantify the latent period of the response and the relation between the intensity of stimulus and latency, to derive action spectra for the photosensor and to investigate "facilitation" by flicker-frequency studies. In this way we can gain much information about the physiology of the photosensor and, indeed, of the whole sensory-motor chain. Although it will be more difficult than investigating the effects of light, we should in time be able to determine the parameters within which chemosensory mechanisms operate also.

H. D. Chapman

Have you attempted to map the distribution of the sense endings in the miracidium?

B. E. Brooker

Yes, but only in the case of *Schistosoma mansoni.*

P. J. Whitfield

Is anything known about the central connections of these supposed sensory cells? If direct electrophysiological recording from digenean receptor cells is technically extremely difficult at the moment is there any possibility of using micro-ablation methods to get information about the modalities of the receptors?

B. E. Brooker

As far as I am aware, no one has yet attempted to determine the central connections of any receptor type. I think micro-ablation would be possible only with those structures such as the eyespots and lateral papillae which are easy to locate by light microscopy. However, you will appreciate that the very small ciliated receptors, of which at least a dozen are found on the apical papilla alone, would be much more difficult to deal with especially in a larva which is itself almost completely covered by cilia.

ABBREVIATIONS USED IN PLATES

aec	aciliated epidermal cell (intercellular ridge)		
agc	apical gland cell	m	mitochondrion
bb	basal body	mf	membrane folds
c	cilium	n	nucleus
cm	circular muscle	ne	nerve ending
cne	ciliated nerve ending	nm	neuronal microtubules
cp	ciliated pit nerve ending	pc	pigment cell
cpl	ciliated plate	r	retinular cell
lgc	lateral gland cell	rb	rhabdomere
lp	lateral papilla	sd	septate desmosome

EXPLANATION OF PLATES

PLATE 1

A. Longitudinal section through the apical papilla of the miracidium of *Schistosoma mansoni* showing the position of some ciliated nerve endings and the extent of the apical gland cell. ×20,000.

B. A ciliated nerve ending which penetrates the thin lateral extension of the apical gland cell. Note the distal electron dense collar and the small vacuoles containing central deposits of material. ×29,600.

C. Transverse section at the level of the basal bodies of a multi-ciliated nerve ending. These are found between the second and third tier of ciliated plates. The cell is connected to the intercellular ridge by septate desmosomes. Miracidium of *S. mansoni*. ×44,500.

PLATE 2

A. Oblique section of the miracidium of *S. mansoni* which passes through two of the several ciliated pit nerve endings to show their relationship to the lateral gland cell. Note that this gland cell has no permanent opening. ×30,000.

B. Longitudinal section of the lateral papilla and, just anterior to it, a ciliated nerve ending. Note the numerous vesicles and neuronal microtubules. Miracidium of *Diplostomum spathaceum*. ×30,000.

PLATE 3

A. Longitudinal section of the lateral papilla on the miracidium of *S. mansoni*. Only the distal part of the nerve ending is visible in this section. The prominent membrane folds are those which surround the ciliated nerve ending. The lateral limits of the intercellular ridge which covers the lateral papilla are marked by septate desmosomes. ×20,000.

B. One of the several ciliated nerve endings found between the first and second tiers of ciliated plates in the miracidium of *S. mansoni*. The membrane folds of the surrounding aciliated epidermal cell are conspicuous. ×30,000.

C. Longitudinal section of one of the ciliated papillae found on the miracidium of *D. spathaceum*. This micrograph shows some of the basal bodies of the ring of cilia. ×22,000.

PLATE 4

A. Section through the ciliated papilla of the miracidium of *D. spathaceum* showing its penetration of an aciliated epidermal cell. The cilia lie roughly parallel to the epidermal cells and their basal bodies possess fibrous rootlet structures. ×29,600.

B. Longitudinal section through the anterior end of the miracidium of *D. spathaceum* showing the position and structure of one of the eyespots. The pigment cell contains two retinular cells. ×9000.

PLATE 5

The eyespot of *D. spathaceum*. Only one retinular cell is visible in this section. Note the closely packed mitochondria and the rhabdomeric microvilli. The position of the cilium derived photoreceptor (top left) relative to that of the eyespot is also shown. ×15,500.

PLATE 6

A. The fifth median rhabdomere of the miracidium of *D. spathaceum*. Note that the mitochondria are dispersed and that the microvilli are randomly orientated. The microvilli are covered by a thin cytoplasmic extension of the left pigment cell (arrow). ×16,000.

B. Section through the cilium derived photoreceptor cell. This section passes through one end of the internally ciliated vacuole which occupies a diverticulum of the cell. Miracidium of *D. spathaceum*. ×30,000.

PLATE 7

A. Transverse section through the ciliated vacuole of the photoreceptor in the miracidium of *S. mansoni*. ×29,500.

B. The cilium derived photoreceptor of the miracidium of *D. spathaceum*. Note that the ciliated vacuole is surrounded by a thin cytoplasmic extension of the cell. ×30,000.

Plate 1

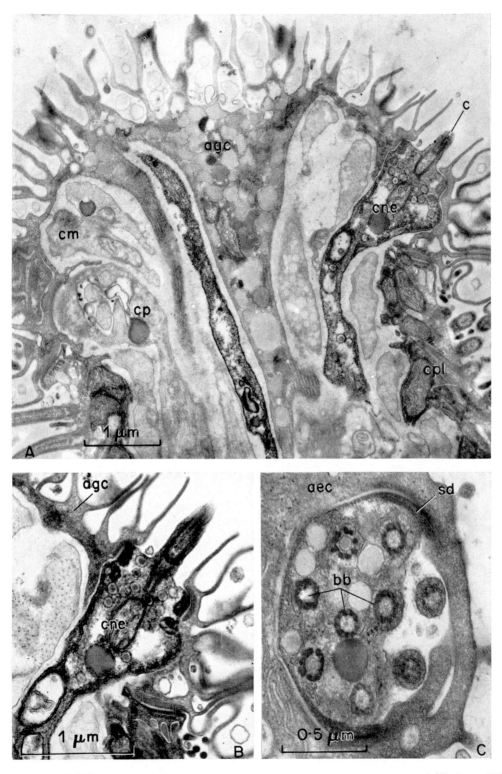

B. E. BROOKER

Plate 2

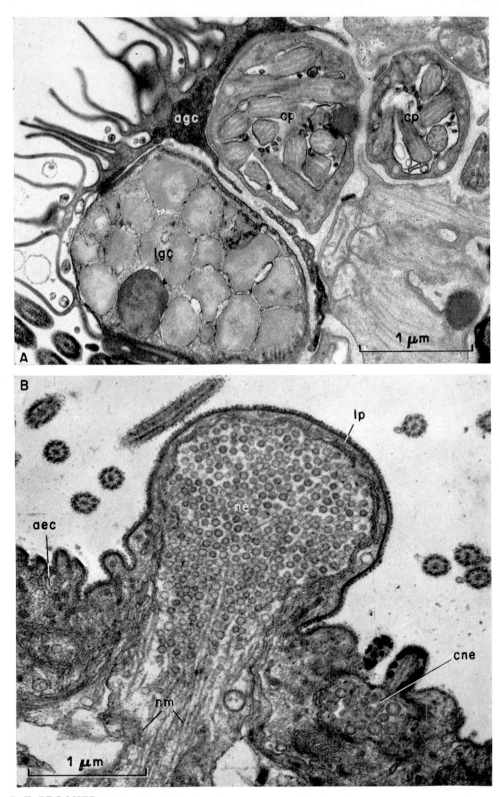

B. E. BROOKER

Plate 3

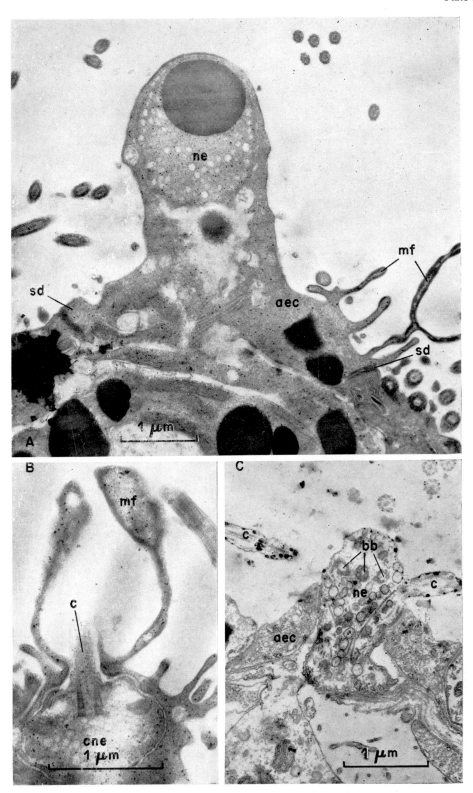

B. E. BROOKER

Plate 4

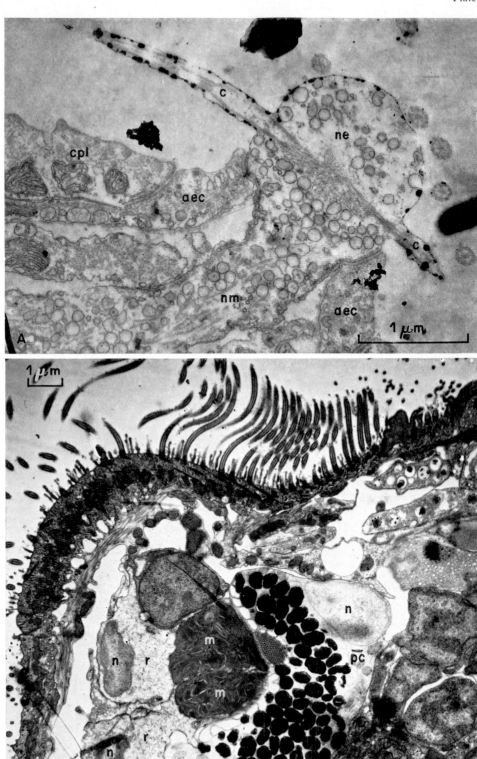

B. E. BROOKER

Plate 5

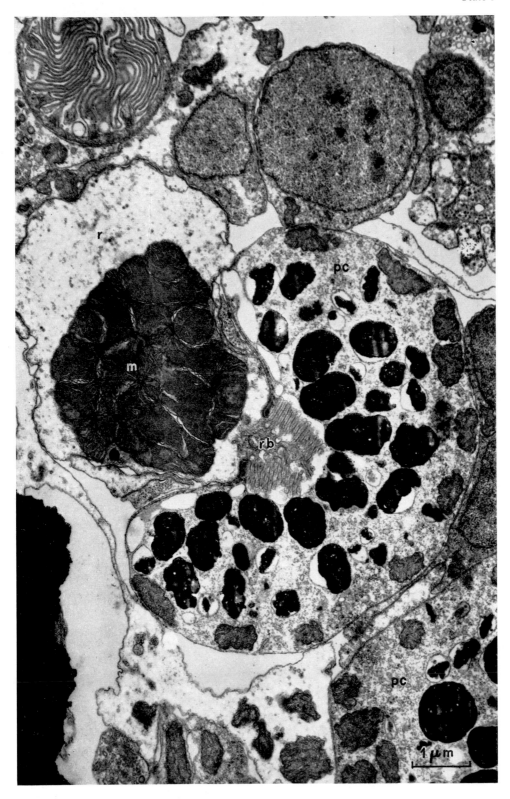

B. E. BROOKER

Plate 6

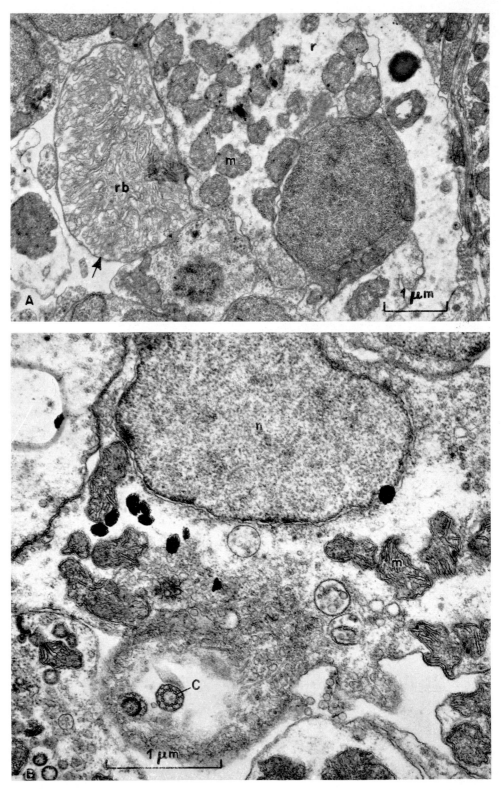

B. E. BROOKER

Plate 7

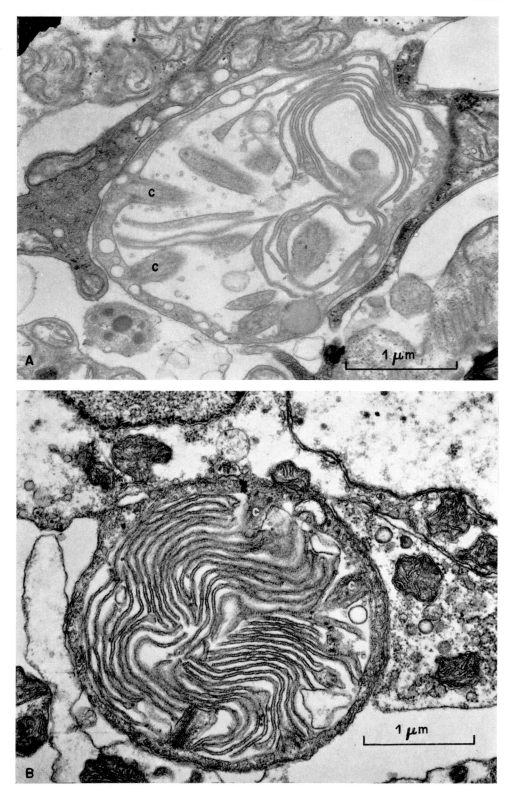

B. E. BROOKER

Sense organs of monogeneans

KATHLEEN M. LYONS

Department of Zoology, King's College, London

A review is given of the kinds of presumed sense organ so far described in monogeneans together with previously unpublished, supplementary information where this has become available. An attempt has been made to interpret the possible functional significance of these "sense organs" and in the absence of electrophysiological or behavioural work this has been done using information on the fine structure and disposition of the "sensilla". The types of organ described are the pigmented eyespots, single receptors ending in a 9 + 2 cilium, compound (uniciliate) receptors consisting of a number of associated nerve terminals each ending in a single cilium and compound (multiciliate) receptors consisting of nerve terminals bearing many cilia. In addition a new kind of internal "sensillum" containing ciliary structures associated with lamellate bodies has been described from *Entobdella soleae* oncomiracidium. A pair of these "sense organs" lies above the eyespots on each side of the head of the larva and it is suggested that they may be ciliary photoreceptors. A preliminary report of another previously undescribed "sense organ" from the haptor of adult *Entobdella soleae* is also given. The application of various techniques which might prove useful in mapping the distribution of the sensilla over the body surface and of tracing the innervation of sense organs by staining the nervous system has been discussed.

CONTENTS

INTRODUCTION

Although there is comparatively little detailed information on the sense organs of monogeneans, scattered accounts in the literature of observations made mainly at the light microscope level suggest that these largely ectoparasitic platyhelminths are fairly richly endowed with sensory structures. The following account reviews the kinds of sense organs so far described in the Monogenea, includes supplementary information on these structures from unpublished results and gives the first description of a new kind of sense organ from the head of the oncomiracidium of *Entobdella soleae.*

MONOGENEAN EYES

Many monogeneans have eyes, although these tend to be more conspicuous in the free living oncomiracidial stage and are often reduced or become absent in the adult parasites. Bychowsky (1957) gives a general account of the kinds of eyes found in monogeneans and more detailed reviews of the arrangement and types of eyes in larval monogeneans have been made by Llewellyn (1963) and by Baer & Euzet (in Grassé, 1961). Where eyes are present these are usually paired and lie between the brain and dorsal body surface. The oncomiracidia of most monopisthocotylineans have four eyespots which are each equipped with a so-called crystalline lens and have a characteristic arrangement whereby a smaller anterior pair is directed posterolaterally and a larger posterior pair is directed anterolaterally (Llewellyn, 1963). This is probably the primitive condition in monogeneans. Most polyopisthocotylineans have only a single pair of eyes which either lack lenses or are equipped with oil droplet lenses, as in *Plectanocotyle gurnardi* (see Llewellyn, 1957a). The eyespots of monopisthocotylineans may persist into the adult but often in a reduced form with the lenses being lost, as occurs in *Entobdella soleae* and *Trochopus pini* (Kearn, 1971). The eyespots of polyopisthocotylineans do not usually persist into the adult (Llewellyn, 1963). The larvae of some monogeneans have a single eyespot, but these probably represent a fused pair of eyespots and may often be equipped with two lenses (Bychowsky, 1957).

The eyes of monogeneans consist of a pigment cup which encloses a sensory cell or cells and, as mentioned, a lens may be present over the opening of the pigment cup. The four eyes of the oncomiracidium of *Entobdella soleae* have been observed to be capable of small movements and may be equipped with muscles (Kearn, pers. comm.; Lyons, unpubl. results) but this has not yet been studied in detail. So far no account of the fine structure of the monogenean eye has been published, but Kearn (pers. comm.) and Lyons have found, using the electron microscope, that the larval eyes of *Entobdella soleae* are rhabdomeric, consisting of sensory cells ending in more or less parallel microvillar stacks enclosed in a cup-shaped pigment cell. Detailed work on the eyes of *Entobdella soleae* is at present in progress. The eyes of monogeneans would seem from these preliminary observations to resemble those found in both the free living platyhelminths (MacRae, 1966) and in larval digeneans (Isseroff & Cable, 1968). The pigment in the eyes of certain echinostome cercariae is melanin (Nadakal, 1960) and this may also be true of monogenean eye cup pigment. In the larva of *Capsala martinieri* extensive areas of pigment resembling that in the four eye cups occurs and is arranged in a bilaterally symmetrical pattern (Kearn, 1963); the significance of this is not known.

The function of eyes in monogeneans

Little is known about the behaviour of monogenean larvae and in particular about the behaviour of larval monogeneans in response to light. Bovet (1967) has noticed that unhatched oncomiracidia of *Diplozoon paradoxum* are activated to escape from the egg by light, although light is not essential for hatching. During their free swimming existence some monogenean larvae are initially positively phototactic but later lose their attraction to light (Bychowsky, 1957; Bovet, 1967). The significance of this initial attraction to light and of the subsequent loss of response to illumination is difficult to assess in a general way since the functions of each behavioural phase would be expected to differ in specific cases. Where the initial phase of positive phototaxis is short-lived (it may last for only two or three minutes in some dactylogyrids (Bychowsky, 1957)) this could merely be a prolongation of a hatching stimulus involving directional activation. Where there is a longer photopositive phase this could act as a distributory device causing the larvae to swim upwards where they may be spread by water currents or, in cases where a pelagic surface feeding fish is used as a host, might bring the larvae into the host area. The loss of attraction to light might be expected in monogeneans parasitizing bottom living fishes. Even if the photopositive phase is not actually succeeded by a photonegative phase or a period of shadow sensitivity (these have not been found), loss of positive phototaxis would be expected, at least when the host area had been located, so that *host specific* clues such as chemosensitivity can dominate.

The pigment eye cups of larval monogeneans, and indeed of larval digeneans, open dorsolaterally so that there is a bilateral rather than frontal light acceptance. This presumably allows a photopositive larva swimming with the dorsal surface uppermost and not spiralling to home upon a frontal light path by simultaneous comparison of the amount of light striking each eye or set of eyes, course correction occurring to keep the amount of light received on each side of the head constant. The dorsal situation of the eyes may give the larva information about which way up it is swimming during the day since the natural biotope consists of illuminated upper waters and dark lower waters. Most monogenean larvae do not spiral as do some digenean miracidia, but swim dorsal side up.

It seems that possession of two pairs of eyespots with an anterior posterolaterally orientated pair and a posterior anterolaterally orientated pair, as found in monopisthocotylineans, may be the primitive condition in monogeneans. This arrangement increases the acceptance angle for light on each side of the body and may enhance directional sensitivity to light, the anterior eyes monitoring light from a posterior direction and the posterior eye monitoring anterior illumination. Separation of information about the direction of light source might be useful to a larva performing three dimensional searching loop or circus movements during a generally photopositive phase. Stimulation of one pair of eyes would occur in both a face and face-away position and this might allow rapid realignment with a light source after an interrupted swimming period that is disorientated with respect to light. Separation into anteriorly orientated and posteriorly orientated receptor cells within the single pair of eyespots of the miracidium *Philopthalmus megalurus* has been noted by Isseroff (1964) and this may be a device for giving the same kind of directional sensitivity to light where only a pair of eyespots is present. Nothing has yet been published about the arrangement of receptor cells in the eyes of a monogenean.

A new sense organ, possibly a ciliary photoreceptor, from the head of
Entobdella soleae *oncomiracidium*

Recent electron microscope observations on the head of the oncomiracidium larva of *Entobdella soleae* using methods described previously (Lyons, 1969a, b) have revealed a new kind of presumed sense organ which may be a ciliary eye. The following description constitutes the first published account of this "sense organ".

Paired round to oval shaped structures, measuring 6 μm (4.8-8 μm) by 4 μm (3.6-4.4 μm) and containing cilia have been found on each side of the head of the oncomiracidium about 7 μm above the eyes. It is possible that these organs correspond to the indoxyl positive regions lying laterally and above the brain in whole mounts of oncomiracidia in which the nervous system has been stained using the indoxyl acetate technique. These regions are indicated in Plate 3A, B. Since there is a further lateral indoxyl positive region posterior to the pigmented eyes (Plate 3A, B) it may be in fact that two pairs of ciliary organs are present; so far only an anterior pair has been located using the electron microscope. The anterior ciliary organ lies about 2 μm under the inner larval epidermis, is not associated with pigment of any kind and does not appear to open to the outside. It is an intracellular structure and contains a central cavity lined by a thin rim of "nerve" cytoplasm (Fig. 1; Plate 1). The peripheral cytoplasm contains basal bodies of aberrant 9 + 2 cilia which project for a short distance into the central lumen. The basal regions of up to eighteen cilia have

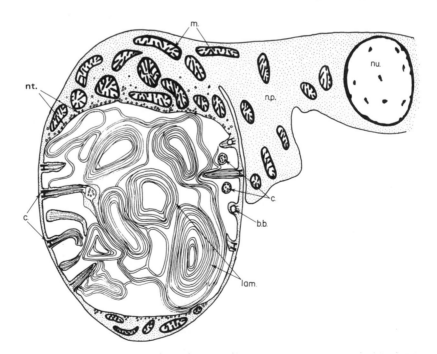

Figure 1. Diagram of the ciliary, suggested photoreceptor from the head of *Entobdella soleae* oncomiracidium. The ciliary organ appears to consist of a single cell with a thick rim of cytoplasm containing mitochondria and (nerve) vesicles at each pole. Laterally the organ is bounded by a thinner cytoplasmic layer containing ciliary basal bodies. The 9 + 2 cilia project into the lumen of the organ and are associated with and may give rise to lamellate bodies.

been seen in sections and these cilia have no conspicuous rootlet and contain outer doublets only for a very short portion of their length, these becoming single microtubules further along the cilium (Plate 2B). The lumen of the organ contains lamellate bodies measuring 1 μm (0.8-1.8 μm) across composed of spiralling intracellular layers bound with unit membranes on both sides (Plates 1 and 2A, C). Up to thirteen layers may be present in each lamellar body. There are indications that the lamellae are extensions of the outer membranes of the cilia (Plate 2C) but this is not absolutely clear and lamellae may also derive from the general lining of the cavity. An extracellular space is present at the centre of many of the lamellar bodies (Plates 1 and 2C). The rim of cytoplasm in which the ciliary basal bodies are embedded contains many microtubules. At the anterior and posterior regions of the organ the cytoplasmic wall is considerably thickened and in addition to the microtubules also contains a profuse number of mitochondria which are packed closely together and have well developed cristae (Plates 1 and 2A). The anterior region of the organ is continuous with a cell body containing a nucleus, scattered ribosomes and many mitochondria (Fig. 1). It has not yet been possible to trace a "nervous" connection any further back than this.

The most likely interpretation of the function of the ciliary organs in *E. soleae* oncomiracidium is that they may be photoreceptors. Unfortunately this supposition will be very difficult to investigate since the ciliary lamellate organs are very small (6 × 4 μm) and lie very close (7 μm) to the brain which presents problems with regard to the possibility of making electrophysiological recordings. The suggestion that these organs may be photoreceptors is based on their considerable similarity to the ciliary eyes of the molluscs *Pecten irradians* (see Miller, 1958), *Cardium edule* (see Barber & Wright, 1969) and *Onithochiton neglectus* (see Boyle, 1969) as well as the feeding tentacle eyes of the annelid *Branchiomma vesiculosum* as described by Krasne & Lawrence (1966). In addition Horridge (1965) has found very similar structures in a ctenophore *Pleurobrachia pileus* which he presumed were photoreceptive on purely morphological grounds. In all of these "photoreceptors" whorling lamellae are associated with and appear to arise from cilia, although there is variation in the structure of the ciliary region, some cilia possessing a central pair of filaments as in the organ of *Entobdella* oncomiracidium (e.g. the lamellate receptors in *Onithochiton* described by Boyle (1969)), others having 9 + 0 cilia (e.g. the presumed photoreceptor in *Pleurobrachia* described by Horridge (1964)). Paired intracellular structures with a single innervation process, containing 9 + 0 cilia have been described from the head of *Fasciola* miracidium by Wilson (1970). These are termed internal club-shaped endings and Wilson considered the possibility that they might be gravity receptors but states that this is not a very likely supposition. It seems possible that the internal club-shaped endings in *Fasciola* miracidium, which has a pair of pigmented eyespots, may be equivalent to the ciliary lamellate head organs of *Entobdella* oncomiracidium and may be photoreceptors. Membrane proliferation is a common feature of most animal photoreceptors (see Moody, 1964) and presumably occurs to accommodate visual pigment. In the rhabdomeric eyes of invertebrates, such as those of larval digeneans (Isseroff & Cable, 1968) membrane expansion takes the form of microvillar extensions of the receptor surface. The membrane extensions provided by the lamellate whorls in ciliary eyes presumably represent a different way of expanding the surface area to accommodate visual pigment. It was once believed that members of different invertebrate groups possessed either a rhabdomeric or a ciliary type of receptor but a number of exceptions have been found. The presence of two kinds of retinal element, one rhabdomeric and one ciliary in the eye of *Pecten*, is well documented (Land, 1968); rhabdomeric and ciliary photoreceptive elements also appear to be present in the dorsal ocelli of the mollusc *Onithochiton neglectus* where a central microvillous rhabdom is surrounded by cells containing lamellae derived from the ends of 9 + 2 cilia (Boyle, 1969). The question then arises as to why a larval monogenean, and possibly a larval digenean, which already possess well developed rhabdomeric pigment cup eyes, should require what is possibly a different kind of eye, a ciliary eye? The most obvious explanation of this would be that the ciliary "eyes" might respond in a different way to illumination and give different information from the rhabdomeric eyes. In *Pecten* eye, the proximal rhabdomeric photoreceptors signal the *presence* of light by firing off nerve impulses at an increased rate when illuminated, whereas the distal ciliary eyes signal the end of illumination by a burst of discharges (the primary "off" response) (see Land, 1968). In the ciliary eye there is a low resting discharge during darkness which stops upon illumination and resumes at a temporarily increased rate at the end of illumination. The ciliary eyes of *Pecten* seem then to be signalling the end of illumination, a function associated with protective shadow responses. The ciliary, siphonal eyes of *Cardium* also produce "off" responses to light (cited in Barber & Wright, 1969) and seem to mediate a shadow response and there is some evidence that the ciliary, feeding tentacle eyes of the annelid *Branchiomma* also mediate a withdrawal response to shadow. As Barber & Wright (1969) point out, it is perhaps unlikely that all ciliated invertebrate photoreceptors produce "off" responses and signal shadowing, so whilst this possibility can be borne in mind as a function of the "ciliary eye" in *Entobdella* oncomiracidium, it cannot exclude other possibilities such as that it may be a long term light receptor, setting, for instance, some hypothetical diurnal or nocturnal activity phase.

SENSE ORGANS OTHER THAN EYES

Presumed sense organs other than eyes have been described at the light microscope level by several authors. "Sensory" structures arising from a projection of the outer surface and ending in a terminal "hair-like" process (presumably a cilium) were found on the body of living *Acolpenteron uretoecetes* by Fischthal & Allison (1940) and on the haptoral suckers of living *Polystomoides renschi* by Rohde (1968). A "receptor" said to have a terminal "hair-like" process and several short lateral processes was described from *Sphyranura osleri* by Wright & MacCallum (1887). The three ciliated tufts present on the heads of the oncomiracidia of *Diplozoon paradoxum* and *Cyclocotyla belones* may also be sense organs since they are not deciduous and remain after the ciliated epidermal cells of these larvae have been shed (cited in Llewellyn, 1963). Silver nitrate staining has successfully been used to locate monogenean "sensory" endings at the light microscope level and might prove a useful method for studying the distribution of sense organs, particularly in fresh water monogeneans and their larvae since in marine forms sodium chloride either has to be substituted for or carefully washed out before silver stains can be applied. Chapman (1968) has described a method that substitutes magnesium nitrate for sodium chloride and involves washing tissue in 0.36M $Mg(NO_3)_2$ for five minutes prior to silver staining. Chapman (1968) used this method prior to staining with silver-osmium for the epidermal cell boundaries of coelenterates, but presumably the method might work prior to silver nitrate staining for "sensilla" in monogeneans. Supposed sensory papillae have been mapped on the oncomiracidium of *Discocotyle sagittata* by Owen (1970) using 0.5% silver nitrate stain of Lynch (1933). Silver stained papillae were found to be mainly anterior and ventral although a pair of "sensilla" was found on the lower side of the lobes supporting the posterior marginal pair of hooks on the haptor. Combes (1968) has identified and plotted the distributions of papillae bearing a fine process (presumably a cilium or cilia) on the bodies of three species of *Polystoma* oncomiracidia after staining with the 1% silver nitrate method of Ginetsinskya & Dobrovolski (1963). The oncomiracidia studied were those of *Polystoma pelobatis, P. integerrimum* and *P. gallieni.* It was noted that some of the "sensilla" were larger than others, that their arrangement was bilaterally symmetrical and that the patterning of the "sensilla" seemed to be constant for oncomiracidia of a particular species (Combes, 1968). A silver staining technique has also been used by Rohde (1968) to locate bulb-like endings and what is tentatively described as a sub-epidermal nerve plexus with free nerve endings in the haptor of adult *Polystoma renschi.* As Wilson (1970), working on the miracidium of *Fasciola hepatica* points out, silver staining methods have to be interpreted with caution since unless light microscope work using silver stains is paralleled with electron microscope work (like that of Chapman & Wilson (1970) on the cercaria of *Himasthla secunda*) there can be no certainty that all structures taking up silver are in fact nerve endings, or conversely that all specialized nerve endings do take up the silver stain. Wilson (1970) found, for example, that certain internal, supposed sensory endings in the miracidium of *Fasciola hepatica* did not stain using silver nitrate methods whereas more superficial "receptors" were stained. Furthermore, silver staining does not distinguish between uniciliate and multiciliate sensilla.

Cholinesterase has been located in the nervous systems of adult *Diplozoon paradoxum* by Halton & Jennings (1964) and *Diclidophora merlangi* by Halton & Morris (1969) who used the indoxyl acetate and acetyl- and butyryl-thiocholine iodide methods controlled with eserine sulphate. Free nerve endings in close proximity to the epidermis were seen in the light microscope in *Diclidophora merlangi* after staining for cholinesterases, but no specialized "sensory" structures were resolved at the light microscope level (Halton & Morris, 1969). The indoxyl acetate method controlled by incubation in 10^{-4} eserine for cholinesterases has also been used as a general stain for the nervous systems of *Gyrodactylus* sp. (see Lyons, 1969a) and the nervous system of the oncomiracidia and adults of both *Entobdella soleae* and *Acanthocotyle lobianchi* (Lyons, unpubl. obs.). The indoxyl acetate stained nervous system of the oncomiracidium of *Entobdella soleae* is shown in Plate 3A, B. The presence of cholinesterase in the nervous systems of these monogeneans implies that choline is one neuro-humour present but there has so far been no attempt to locate any other transmitter substance in the nervous system of a monogenean. Dopamines, noradrenaline and serotonin have been found in planarian nervous systems (see Welsh & King, 1970) so it is quite possible that transmitters other than cholinesterase are important in the parasitic groups. Granules measuring 0.09-0.1 μm which may be neurosecretory have been located in the nerves supplying the specialized spike sensillum of *Gyrodactylus* sp. (see Lyons, 1969b).

Electron microscope studies

At the electron microscope level, three main kinds of presumed sense organ, all of which end in cilia and open through the epidermis, have been described so far. These are listed below:

(1) *Single receptors* consisting of a nerve bulb and terminal, non-motile 9 + 2 cilium with no rootlet (Lyons, 1969a; Halton & Morris, 1969).
(2) *Compound (uniciliate) receptors* consisting of a number of associated nerve terminals each of which bears only a single terminal cilium (see Lyons, 1969a, b).
(3) *Compound (multiciliate) receptors* made up of one or a few nerve endings only, each of which bears many modified cilia (Lyons, 1969b).

There is considerable variation in the detailed structure of the sensilla of each class, for instance compound uniciliate receptors may end in 9 + 2 cilia or modified cilia, rootlets may be present or absent etc. A more detailed account of these different kinds of sense organs is presented and the possible functional significance of some of these receptors discussed. It must be emphasized that a sensory function is attributed to these structures only on comparative morphological grounds; there is no neurophysiological or behavioural evidence to condone this assumption at present.

Single receptors

These are probably the most frequently observed "sensilla" in monogeneans and consist of a nerve bulb sealed into the living epidermis by septate desmosomes and equipped with a single, non-motile, 9 + 2 cilium which may be up to 9-11 μm in length. The structure of the single receptors of *Gyrodactylus* sp. and adult *Entobdella soleae* has been described in detail by Lyons (1969a). The terminal cilium of these receptors has a normal basal body but no basal foot or rootlet. Nine transitional fibres radiate outwards from the basal body. The nerve cytoplasm in which the cilium is embedded contains vesicles, neurotubules and mitochondria whilst the periphery of the nerve terminal is strengthened by a dense collar. Further concentric dense zones lie outside the nerve bulb in the epidermal cytoplasm (Fig. 2A, B). This kind of receptor has

Figure 2. Diagram showing a variety of different, independently innervated sensilla from various monogeneans. A. Single receptor from *Gyrodactylus* sp. as seen with the light microscope. B. Single receptor as seen with the electron microscope. C. Previously undescribed peg organ from the body of *Acanthocotyle lobianchi* which terminates in a single cilium and is connected by a nerve to a bipolar neurone. D. Compound (uniciliate) receptors from the head of adult *Entobdella soleae*. E. Compound (uniciliate) receptors from the head of *Entobdella oncomiracidium*. F. Spike sensillum from the head of *Gyrodactylus* sp. showing striped ciliary rootlets and "neurosecretory" granules in the nerve processes.

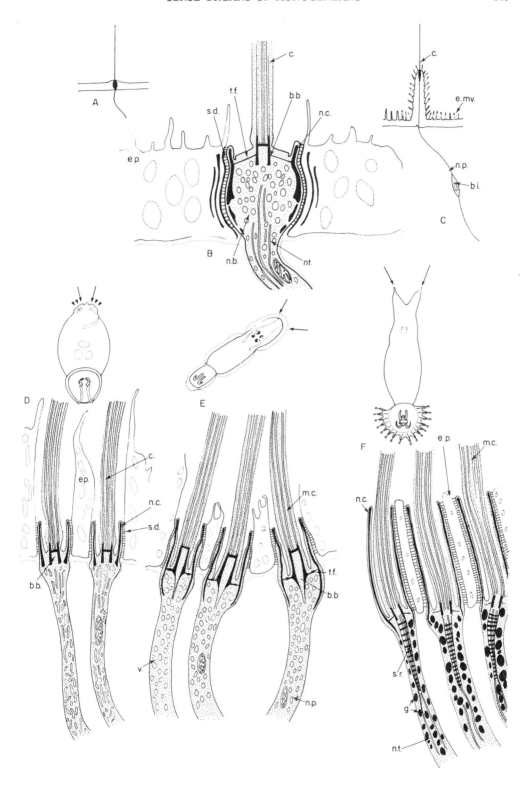

also been found in the oncomiracidium of *Entobdella soleae,* in the juvenile stage of *Amphibdella flavolineata* and in *Leptocotyle minor* (see Lyons, 1969a). Similar sensilla found in the polyopisthocotylinean *Diclidophora merlangi* by Halton & Morris (1969) have a nerve bulb strengthened by dense ring-like structures rather than by a complete "cylindrical" nerve collar. The endoparasitic digeneans and cestodes also have sensilla of this type (see Lyons, 1969a and Chapman & Wilson, 1970 for a review). Matricon-Gondran (1971) has described single receptor-type sense organs in digenean larvae which have ciliary rootlets or basal fibres. It has been suggested that these organs are tangoreceptors (see Lyons, 1969a) or in the case of ensheathed sensilla found in the digenean *Schistosoma mansoni,* possibly rheoreceptive (Morris & Threadgold, 1967). These suggestions are based partly on the distribution of the endings over the body and partly on morphological similarity to proven mechanoreceptors in other invertebrates (see Horridge, 1965). If the single receptors are mechanoreceptive it is of course quite possible that they could act as both touch receptors and rheoreceptors. They may well have different functions in different worms and may also vary in sensitivity at different stages of the parasite's life cycle. Tangoreceptors would obviously be of use to monogeneans which have first to make contact in the larval stage with the host and then to maintain contact by means of the anterior glands and posterior haptor. Several skin parasitic monogeneans, e.g. *Gyrodactylus* sp. migrate over the surface of their hosts by alternately attaching and detaching anterior and posterior adhesive organs so that contact receptors on the head and near the haptor, as are the single receptors of *Gyrodactylus* sp., might provide useful proprioceptive information during this kind of behaviour. Contact receptors of some kind are presumably used during feeding and in copulation or in the exchange and handling of spermatophores as occurs, for instance, in *Entobdella soleae* (Kearn, 1970). Some gill parasitic monogeneans, e.g. *Kuhnia scombri,* tend to occupy relatively sheltered positions in the branchial chamber of the host fishes where they are not subjected to the full force of the gill ventilating current (Llewellyn, 1957b), in this case it might be envisaged that possession of rheoreceptors would be of some use. Comparatively little is known about the innervation of single receptors. Lyons (1969a) described a direct connection from the nerve bulb of a single receptor in *Gyrodactylus* sp. to the ventral nerve cord and found no cell body specifically associated with this receptor, but this may have been an artefact due to coarse indoxyl deposition and needs reinvestigation. A new, so far undescribed, single receptor on *Acanthocotyle lobianchi* consisting of a finger-like projection of the epidermis 7-10 μm long and containing a single cilium which projects for a further 2-5 μm beyond the tip of the projection has been shown by methylene blue staining to be connected via a thin nerve process 4 μm long to a bipolar neurone 12 μm in length (Fig. 2C) (Lyons, unpubl. results). To stain with methylene blue 15 drops of a 0.5% stock solution of methylene blue were added to sea water containing living worms, or to 50 : 50 7.5% $MgCl_2$:sea-water. Animals were fixed in 3% glutaraldehyde for 30 min and then in 10% ammonium molybdate before clearing through xylol and mounting in Damnar xylol (Alexandrowicz, pers. comm.). The peg-like organs are easily visible with the light microscope and are extremely numerous (Plate 3C). They occur over the whole body on both dorsal and ventral surfaces, but are concentrated on the head and in a region just anterior to the psuedohaptor where they are mainly lateral in position. They also occur over the muscular lips around the common genital atrium. It is proposed to describe these organs in detail elsewhere at a later date.

Compound (uniciliate) receptors

These consist of a number of associated nerve endings each ending in a single cilium. The periphery of each nerve terminal is thrown up into a ridge-like nerve collar which contains a dense internal thickening and each nerve collar is sealed into the epidermis by means of septate desmosomes (Lyons, 1969a, b). The precise innervation of these organs needs to be studied since although this type of sensillum seems to be composed of separate endings at epidermal level, it is quite possible that the processes may unite on to a single nerve cell further back towards the brain. Several different kinds of cilium and associated structures have been described from the three kinds of compound uniciliate receptor known.

(a) The head organs of adult *Entobdella soleae.* These occur in front of the adhesive head lappets and terminate in an undetermined number of 9 + 2 cilia which have a basal plate and a basal body associated with nine transitional fibres but no rootlet. Electron lucent tubular "vesicles", neurotubules and mitochondria occur in the nerve cytoplasm (Fig. 2D). The position of these compound sense organs on the head of adult *Entobdella* makes either a chemoreceptive or a tangoreceptive role equally likely.

(b) Grouped receptors on the head of *E. soleae* oncomiracidium. These were originally referred to as cone sensilla (Lyons, 1969b) but there is little evidence that these receptors correspond to conical projections seen with the light microscope, so it is proposed to term these organs grouped receptors in future. These sensilla occur ventrally next to the gland duct openings on each side of the head of the oncomiracidium (Lyons, unpubl. results). Each receptor consists of seven to ten nerve terminals each of which is equipped with a highly modified cilium which contains peripheral but disorganized doublets in addition to many supernumerary microtubules. There is a basal plate and the basal body is associated with transitional fibres which radiate outwards to make contact with the cylindrical internal thickening of the nerve collar. The cilia have a short rootlet which tapers off from the basal body (not previously described). Large clear vesicles are characteristic inclusions of the nerve endings to this organ (Fig. 2E). It is not possible to ascribe a function to these sensilla as they exist along with a whole bank of presumed receptors. These include the ciliated pits (see p. 192) but also so far undescribed organs such as a nerve ending or endings terminating in a macrocilium packed with single microtubules and what appears to be a closed bulb lined with basal body-like structures which are orientated predominantly towards the centre of the bulb and have stumpy, inflated microtubule-containing ends (Lyons, unpubl. results). This structure is quite different from the ciliary lamellate eye-like receptor described earlier (p. 184).

(c) The spike sensilla of *Gyrodactylus* sp. These have already been described from both scanning and transmission electron microscope observations (Lyons, 1969b). The spike sensilla are situated on the anteriormost tips of the head "horns" of *Gyrodactylus* and lie dorsal to the adhesive gland openings. They are flanked by four or five single receptors, as already described. The spike sensillum consists of ten nerve endings each bearing a modified cilium 11 μm long which has a prominent striped rootlet that may be continuous with the many neurotubules present in the cytoplasm of the nerve terminal (Fig. 2F). The cilia are sheathed in epidermis for much of their length but the sheath is open at the tip. Dense inclusions 0.09-0.1 μm in diameter occur in the cytoplasm of the receptor terminals and may be neurosecretory material. Since these spike sensilla are surrounded by "tangoreceptors" it has been argued that these head organs may be chemoreceptors (Lyons, 1969b).

Compound (multiciliate) receptors

This kind of receptor is usually made up of one to five nerve endings only, each of which is equipped with many highly modified cilia. The edges of each multiciliate terminal are formed into a nerve collar, there is no nerve collar around the individual cilia in these receptors. Two kinds of multiciliate receptor have already been described (Lyons, 1969b) and a brief synopsis of each is given below, together with supplementary (unpubl.) information where this has become available.

(a) Ciliated pits of *Entobdella soleae* oncomiracidium. Recent observations have shown that there is a group of three ciliated pits situated ventrally on each side of the head near the adhesive glands, one of the pits being smaller (1μm) than the other two (2 μm). The larger ciliated pits each contain a nerve ending bearing 15-18 highly modified cilia (Lyons, unpubl. results). The nature of the basal body of each cilium and of the nerve inclusions present is not sufficiently well known to deserve comment here (Fig. 3A). The oncomiracidium of *E. soleae* has been shown to distinguish its host from closely related species by means of chemical clues (Kearn, 1967) so one would expect to find chemoreceptors on the head of this larvae. Which of the many "sensilla" being discovered on the head of the larva (p. 188) may act as chemosensors is going to be difficult or impossible to discover without electrophysiological recording.

(b) Sensilla on the head and general body of *Acanthocotyle elegans* and *A. lobianchi*. These were originally described from around the feeding organ of *Acanthocotyle elegans* (see Lyons, 1969b) but in fact are generally distributed near the adhesive regions on the head and over the body margins of the head region. They occur less frequently towards the pseudohaptor and have not been found on the latter. What seem to be identical sensilla have been found in similar regions on *A. lobianchi* (unpubl. results by Lyons), collected from the ventral surface of *Raia clavata* (*Acanthocotyle elegans* occurs on the dorsal surface of this ray). The following description applies to the sensillum from both *A. elegans* and *A. lobianchi* and extends information already given on the sensillum from *A. elegans*. In the light microscope the multiciliate receptors from *Acanthocotyle* correspond with bump-like structures measuring 5 μm across and 3μm in height, bearing three to five bunches of immobile cilia 5-7 μm in length (Plate 3C). In the electron microscope it is seen that each bunch of cilia arises from a single nerve process and that about 7-10 cilia are present on each of the three to five nerve endings comprising the sensillum. The cilia are modified and contain many microtubules and have a basal body region containing supernumerary microtubules inside and outside the triplet ring

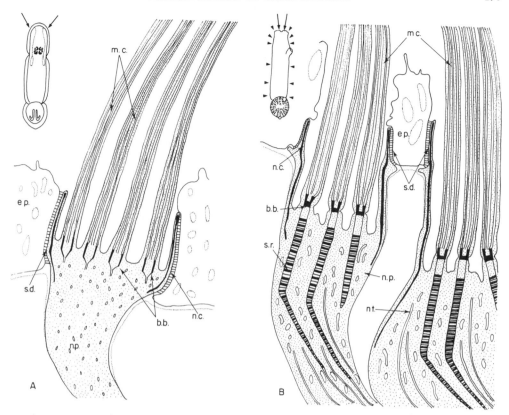

Figure 3. Diagram showing compound (multiciliate) receptors from two monogeneans. A. Ciliated pit on the head of *Entobdella* oncomiracidium. B. Compound multiciliate receptors on the body of *Acanthocotyle lobianchi*, showing the striped rootlets to the cilia.

(Lyons, 1969b). The cilia have a basal plate and there is a long, tapering striped rootlet which bends sharply through about 60° about one-third of the way along its length. The nerve ending contains electron opaque inclusions which often appear tubular in profile (Fig. 3B).

It seems doubtful that these multiciliate receptors would function very well as tangoreceptors. The lack of a 9 + 2 arrangement in the terminal cilia and the fact that three to five bunches of cilia occur on each sensory hillock would certainly imply that these organs may not be very sensitive to directional mechanical stimulation. In view of the fact that peg-like organs (see p. 190) and what are apparently single receptors, both of which may be mechanoreceptors occur in the neighbourhood of these multiciliate receptors (Plate 3C) it is perhaps more likely that the latter are chemosensitive.

The haptoral knobs of adult Entobdella soleae

The ventral surface of the haptor of adult *Entobdella soleae* is covered with rounded papillae or knobs, which measure between 2.5-19 μm across. Stereoscan observations demonstrate that there are about 800 of these structures on the ventral surface of the haptor (Lyons, unpubl. results). Examination of thin sections through the knobs using the transmission electron microscope shows that they are packed with nerves which connect with the main haptoral nervous system and that they may therefore be sensory. No cilium, basal body or rootlet structure is present (Lyons, unpubl. results). It is tentatively suggested that these structures may be mechanoreceptors and possibly that they are contact receptors. A more precise description of these organs and a more searching consideration of their function will be published shortly elsewhere. The absence of a cilium or associated structures from a monogenean presumed sense organ (other than the rhabdomeric eyes) is unusual. Wilson (1970) has described what are possibly aciliate receptors in another platyhelminth; these are the lateral bulbous endings of *Fasciola hepatica* miracidia.

DISCUSSION

The interpretation of the organs described in this paper as "sensory" rests purely upon their morphological resemblance to proven sense organs in other, non parasitic, invertebrates and there is a great need for behavioural investigations and neurophysiological recording to support this contention. Of the presumed sensilla so far observed, the haptoral knobs of adult *Entobdella soleae* (see p. 193) are perhaps the most amenable to study with microelectrodes since they may measure up to 19 μm across; most of the other "sensilla" described are very small and would be difficult to record from. There is also considerable scope for development of methods which will allow the positions of the various sensilla to be mapped more accurately. Most monogeneans are marine and have not been investigated with superficial silver staining techniques that locate sensilla. A method that may make this possible has been outlined on p. 187. Stereoscan methods that use preparatory critical point drying (Boyde & Wood, 1969) and freeze drying (Small & Marszalek, 1969), techniques that prevent sensory cilia being torn off the surface of the specimen by meniscus forces created when thin films of liquids are allowed to evaporate also have considerable application in this field. Far too little is known about the innervation of monogenean sense organs. The nervous systems of several adult monogeneans and two species of monogenean larva (see p. 187 and Plate 3A, B) have been stained using either the indoxyl acetate or acetyl-thiocholine methods for esterases (Halton & Jennings, 1964; Halton & Morris, 1969; Lyons, 1969a) but although these stain the brain and nerve cords they do not always allow visualization of the fine, superficial nerve terminals to sense organs. The methylene blue staining technique of Alexandrowicz (see p. 190) has given promising but variable results for superficial nerves when applied to *Acanthocotyle lobianchi.*

Most of the monogenean "sensilla" so far observed with the electron microscope end in cilia, some of which are highly modified and contain many microtubules. The function of cilia in sensory endings of diverse kinds ranging from mechanoreceptors, such as that described in a chaetognath by Horridge & Boulton (1967), to chemoreceptors, such as the olfactory sense organs of frogs (Reese, 1965) and vertebrate photoreceptors, is far from clear. Thurm (1968) found that the large motor abfrontal cilia on the gills of *Mytilus* showed an inherent mechanosensitivity in that imposed bending of a resting cilium could invoke an active stroke. It was also suggested that the mechanosensitivity could possibly be localized in the basal body including the basal foot (Thurm, 1968).

If ordinary motor cilia possess the ability to translate an applied mechanical stimulus into movement which may involve transduction mediated by the basal body and membrane depolarization, it may be that the basal body of sensory endings can also be implicated in transduction of a stimulus. This is easier to imagine for mechanoreceptors, however, than for those chemoreceptors that have non-motile terminal cilia and for photoreceptors. If indeed the basal body of sensory cilia is the main transductive site, as hinted possible by Thurm (1968) and Laverack (1968) it may be that the outer ciliary shaft, where present, serves to conduct information received at a distance from the nerve ending to the basal body region, i.e. that "it serves only as an extension of the cell into the surrounding environment" (Laverack, 1968). If the arrangement of the supporting and perhaps conducting microtubules within sensory cilia other than polarized mechanoreceptors is of little functional consequences this would explain why the microtubule arrangement within the sensory cilia of monogeneans and other animals is so variable (see Lyons, 1969a, b for review). The cilium endings of those monogenean "sense organs" visible with the light microscope are non-motile sensory cilia; motile sensory cilia have however been described, for instance in the olfactory epithelium of frogs (Reese, 1965) and in the gemmiform pedicellariae of *Echinus* (cited in Laverack, 1968).

A new kind of "sense organ" from the sides of the head of *E. soleae* oncomiracidium has been described in this paper. This organ contains only stumpy basal ciliary shafts; the ends of the cilia are expanded into thin processes which contribute to intracellular lamellate structures. The possibility that this structure may be a photoreceptor has already been discussed (p. 184).

A sense organ has been found in a monogenean that lacks any kind of internal cilium-like structure or rootlet; this is the haptor knob of adult *Entobdella soleae*. A preliminary report of this structure has been given on p. 193. It is tentatively supposed from the position of these organs that they may be mechanoreceptors registering local compression caused by contact with the host. A detailed description of these organs is to be published elsewhere.

ACKNOWLEDGEMENTS

I am indebted to the Browne Fund of the Royal Society for a grant that enabled me to work at the Laboratories of the Marine Biological Society of the United Kingdom, Plymouth, during the course of this work.

REFERENCES

BAER, J. & EUZET, L., 1961. In P. Grassé (Ed.) *Traité de Zoologie, 5*. Paris: Masson et Cie.

BARBER, V. C. & WRIGHT, D. E., 1969. The fine structure of the eye and optic tentacle of the mollusc *Cardium edule. J. Ultrastruct. Res., 26:* 515-528.

BOVET, J., 1967. Contribution à la morphologie et à la biologie de *Diplozoon paradoxum* von Nordmann, 1832. *Bull. Soc. neuchâtel. Sci. nat., 90*(11): 63-159.

BOYDE, A. & WOOD, C., 1969. Preparation of tissues for surface-scanning electron microscopy. *J. Microsc., 90:* 221-249.

BOYLE, P. R., 1969. Fine structure of the eyes of *Onithochiton neglectus* (Mollusca : Polyplacophora). *Z. Zellforsch. microsk. Anat., 102:* 313-332.

BYCHOWSKY, B. E., 1957. *Monogenetic trematodes, their classification and phylogeny.* Moscow, Leningrad: Academy of Sciences, U.S.S.R. (English translation edited by W. J. Hargis, 1961. Washington: American Institute of Biological Sciences.)

CHAPMAN, D. M., 1968. A new type of muscle cell from the subumbrella of *Obelia. J. mar. biol. Ass. U.K., 48:* 667-688.

CHAPMAN, H. D. & WILSON, R. A., 1970. The distribution and fine structure of the integumentary papillae of the cercaria of *Himasthla secunda* (Nicoll). *Parasitology, 61:* 219-227.

COMBES, C., 1968. Biologie, écologie des cycles et biogéographie de digènes et monogènes d'amphibiens dans l'est des Pyrenées. *Mém. Mus. natn. Hist. nat. Paris (Ser. A),* [*51*] 1-195, Fasc. unique.

FISCHTHAL, J. H. & ALLISON, L. N., 1940. *Acolpenteron uretoecetes* n.g., n.sp., a monogenetic trematode from the ureters of black basses. *J. Parasit., 26:* 34-35.

GINETSINSKYA, T. A. & DOBROVOLSKI, A. A., 1963. A new method for finding sensilla in trematode larvae and the significance of these structures in classification. *Dokl. Akad. Nauk. SSSR., 151:* 460-463.

HALTON, D. W. & JENNINGS, J. B., 1964. Demonstration of the nervous system in the monogenetic trematode *Diplozoon paradoxum* Nordmann by the indoxyl acetate method for esterases. *Nature, Lond., 202:* 510-511.

HALTON, D. W. & MORRIS, G. P., 1969. Occurrence of cholinesterase and ciliated sensory structures in a fish gill fluke *Diclidophora merlangi* (Trematoda : Monogenea). *Z. Parasitenk., 33:* 21-30.

HORRIDGE, G. A., 1964. Presumed photoreceptive cilia in a ctenophore. *Q. Jl microsc. Sci., 105:* 311-317.

HORRIDGE, G. A., 1965. Non-motile sensory cilia and neuromuscular junctions in a ctenophore independent effector organ. *Proc. R. Soc. (B), 162:* 333-350.

HORRIDGE, G. A. & BOULTON, P. S., 1967. Prey detection by Chaetognatha via a vibration sense. *Proc. R. Soc. (B), 168:* 413-419.

ISSEROFF, H., 1964. Fine structure of the eyespot in the miracidium of *Philopthalmus megalurus* (Cort, 1914). *J. Parasit., 50:* 549-554.

ISSEROFF, H. & CABLE, R. M., 1968. Fine structure of photoreceptors in larval trematodes. A comparative study. *Z. Zellforsch. microsk. Anat., 86:* 511-534.

KEARN, G. C., 1963. The oncomiracidium of *Capsala martinieri,* a monogenean parasite of the sun fish (*Mola mola*). *Parasitology, 53:* 449-453.

KEARN, G. C., 1967. Experiments on host-finding and host-specificity in the monogenean skin parasite *Entobdella soleae. Parasitology, 57:* 585-605.

KEARN, G. C., 1970. The production, transfer and assimilation of spermatophores by *Entobdella soleae,* a monogenean skin parasite of the common sole. *Parasitology, 60:* 301-311.

KEARN, G. C., 1971. The attachment site, invasion route and larval development of *Trochopus pini,* a monogenean from the gills of *Trigla hirudo. Parasitology, 63:* 513-525.

KRASNE, F. B. & LAWRENCE, P. A., 1966. Structure of the photoreceptors in the compound eyespots of *Branchiomma vesiculosum. J. Cell Sci., 1:* 239-248.

LAND, M. F., 1968. Functional aspects of the optical and retinal organization of the mollusc eye. *Symp. zool. Soc. Lond., 23:* 75-96.

LAVERACK, M. S., 1968. On superficial receptors. *Symp. zool. Soc. Lond., 23:* 299-326.

LLEWELLYN, J., 1957a. The larvae of some monogenetic trematode parasites of Plymouth fishes. *J. mar. biol. Ass. U.K., 36:* 243-259.

LLEWELLYN, J., 1957b. The mechanisms of attachment of *Kuhnia scombri* (Kuhn, 1829) (Trematoda : Monogenea) to the gills of its host *Scomber scombrus* L., including a note on the taxonomy of the parasite. *Parasitology, 47:* 30-39.

LLEWELLYN, J., 1963. Larvae and larval development of monogeneans. *Adv. Parasit., 1:* 287-326.

LYNCH, J. E., 1933. The miracidium of *Heronimus chelydrae* MacCallum.*Q. Jl microsc. Sci., 76:* 13-33.

LYONS, K. M., 1969a. Sense organs of monogenean skin parasites ending in a typical cilium. *Parasitology, 59:* 611-623.

LYONS, K. M., 1969b. Compound sensilla in monogenean skin parasites. *Parasitology, 59:* 625-636.

MACRAE, E. K., 1966. The fine structure of photoreceptors in a marine flatworm. *Z. Zellforsch. mikrosk. Anat., 75:* 469-484.

MATRICON-GONDRAN, M., 1971. Etude ultrastructurale des recepteurs sensoriels tegumentaires de quelques trématodes digénétiques larvaires. *Z. Parasitenk., 35:* 318-333.

MILLER, W. H., 1958. Derivatives of cilia in the distal sense cells of the retina of *Pecten. J. biophys. biochem. Cytol., 4:* 227-228.

MOODY, M. F., 1964. Photoreceptor organelles in animals. *Biol. Rev., 39:* 43-86.

MORRIS, G. P. & THREADGOLD, L. T., 1967. A presumed sensory structure associated with the tegument of *Schistosoma mansoni. J. Parasit., 53:* 537-539.

NADAKAL, A. M., 1960. Chemical nature of cercarial eye-spots and other tissue pigments. *J. Parasit., 46:* 475-483.

OWEN, I. L., 1970. The oncomiracidium of the monogenean *Discocotyle sagitta. Parasitology, 61:* 279-292.

REESE, T. S., 1965. Olfactory cilia in the frog. *J. Cell Biol., 25:* 209-230.

ROHDE, K., 1968. Lichtmikroskopische untersuchungen an den Sinnesrezeptoren der Trematoden. *Z. Parasitenk., 30:* 252-277.

SMALL, E. B. & MARSZALEK, D. S., 1969. Scanning electron microscopy of fixed, frozen and dried potatoes. *Science, N.Y., 163:* 1064-1065.

THURM, U., 1968. Steps in the transducer process of mechanoreceptors. *Symp. zool. Soc. Lond., 23:* 199-216.

WELSH, J. H. & KING, E. C., 1970. Catecholamines in planarians. *Comp. Biochem. Physiol., 36:* 683-688.

WILSON, R. A., 1970. Fine structure of the nervous system and specialised nerve endings in the miracidium of *Fasciola hepatica. Parasitology, 60:* 399-410.

WRIGHT, R. R. & MACCALLUM, A. B., 1887. *Sphyranura osleri,* a contribution to American helminthology. *J. Morph., 1:* 1-48.

DISCUSSION

P. J. Whitfield

It seems that in several invertebrates (e.g. the honey bee and the octopus) which are sensitive to the plane of polarized light, this sensitivity is correlated with the presence of sets of photosensitive lamellae or microvilli which are at precise angles to each other. Your electromicrographs of the eye of *Entobdella* oncomiracidium seem to show a similar differentiation into two sets of photosensitive microvilli at an angle to one another. If this arrangement should prove to be comparable with the arrangements in the honey bee and octopus, could one envisage any possible biological significance for the appreciation of the plane of polarized light in these larvae?

K. M. Lyons

Sensitivity to the plane of polarized light has of course been implicated as a navigational aid and in the possession of a time sense in other invertebrates. However, until more is known about the precise arrangement of the receptor cells in the rhabdomeric eye of monogeneans and digeneans it would seem a little premature to speculate on this interesting idea.

C. Combes

Using silver staining at the light microscope level to localize superficial (presumed) receptors in *Polystoma* oncomiracidia and juveniles I have found that certain of these "receptors" appear to become lost or at least become unstainable after the larvae become attached to their tadpole host. Most of the sense organs that are "lost" subsequent to attachment are situated dorsally and seem to be slightly larger than the other sense organs that are retained, so may have a different function from the latter. Have you noticed any change in the sense organ complement of *Entobdella soleae* larvae after attachment to the host?

K. M. Lyons

As yet I have not been able to plot the distribution of the superficial receptors over the body of this larva at the light microscope level. The heavy epidermal ciliation of the free swimming larvae makes it difficult to locate the ciliary endings of sense organs using phase contrast microscopy and there are problems with regard to using silver stains for nerve endings on this marine material. One would perhaps expect that sense organs other than eyes might regress or change their properties having performed a specific function in one particular phase of the life history. It seems equally probable that adult parasites might develop new kinds of sense organ to monitor the different kinds of conditions encountered in the parasitic phase of their existence. There must certainly be great scope for remodelling the sensory equipment in digeneans where the life cycle includes polyembryony and considerable "metamorphosis" to the adult form.

D. Davenport

These observations of Dr Lyons bear such a close relation to work being carried out in our laboratory as to necessitate a comment. If one were to investigate the photoresponses of the forms which have a pair of eyes faced forward, a pair faced backward and a number of ciliary "photosensors" one would soon realize the utter futility of using the word "phototaxis" for any sort of photoresponse that results in movement toward or away from light. Be that as it may, one most certainly could determine how these eyes work with existing techniques, by studying the angles of course correction under sequential illumination from different directions, as we have already done with photosensitive unicellular flagellates. Certainly the behaviour of such multi-eyed forms will be found to vary in a predictable way dependent upon the direction of stimulation, and this behaviour will probably not vary in the same way as that of a two-eyed control form under similar stimulation. In any case monogenean oncomiracidia with four rhabdomeric eyes must obviously be especially suitable for studies of photo orientation.

K. M. Lyons

Whilst the importance of the kind of experimental approach that Professor Davenport advocates should not be underestimated and provides an elegant way of investigating the behavioural repertoire of organisms to light, I feel that one should not lose sight of the actual environmental conditions under which these parasite larvae operate. In the case of *Entobdella soleae* oncomiracidia, those larvae hatch from eggs stuck to sand grains on the sea bed and must for the most part experience not the thin directional light beams used in experimental set ups but navigate in a diffusely lit environment. They may of course experience directional light in thin beams when swimming between natural objects such as rocks etc. but in general one might suspect that these photosensitive larvae might be making comparatively simple responses to distinguish between a dark bottom field and diffusely illuminated upper waters.

ABBREVIATIONS USED IN FIGURES AND PLATES

b.b.	Basal bodies of cilia		
bi.	Bipolar neurone	m.c.	Modified terminal cilia
c.	Cilia	mt.	Microtubules
cil?	Ciliary organ?	mv.	Microvilli
c.o.	Compound (multiciliate) sensilla	n.b.	Nerve bulb
e.	Pigmented eyes	n.c.	Nerve collar
e.mv.	Epidermal microvilli	n.p.	Nerve process
ep.	Epidermis	nt.	Neurotubule
g.	Granules	nu.	Nucleus of cell body
h.	Larval haptor	p.o.	Peg organ sensillum
l.	Lens	s.d.	Septate desmosome
lam.	Lamellate bodies	s.r.	Striped rootlet to cilium
m.	Mitochondria	t.f.	Transitional fibres

EXPLANATION OF PLATES

PLATE 1

Electron micrograph of a longitudinal section through one of the ciliary organs in the head of *Entobdella soleae* oncomiracidium. The ciliary organ seems to consist of a single cell with a central cavity into which project 9 + 2 cilia. Lamellate structures composed of thin membrane bound whorls of intracellular substance are present in the lumen of the organ. The poles of the cell contain mitochondria. ×27,430.

PLATE 2

A. Electron micrograph of a section through the ciliary head organ of *Entobdella soleae* oncomiracidium showing microtubules and mitochondria in the peripheral cytoplasm and ciliary basal bodies and lamellate structures in the lumen. ×43,840.
B. Transverse section through a cilium from the ciliary organ showing the two central filaments and aberrant outer tubules. ×87,680.
C. Detail of the ciliary organ showing the apparent origin of lamellae from the end of a cilium. ×47,240.

Plate 1

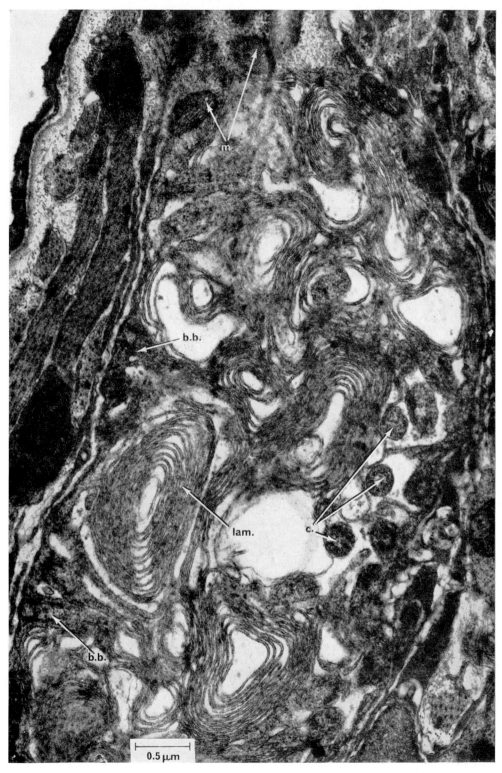

Plate 2

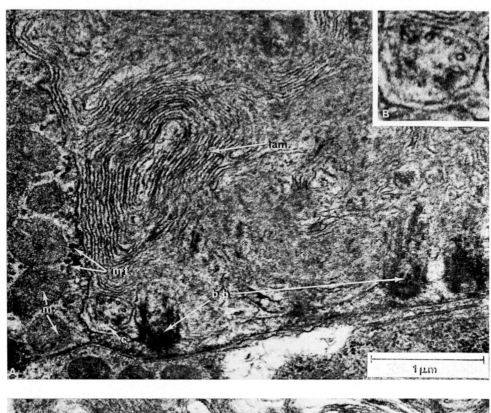

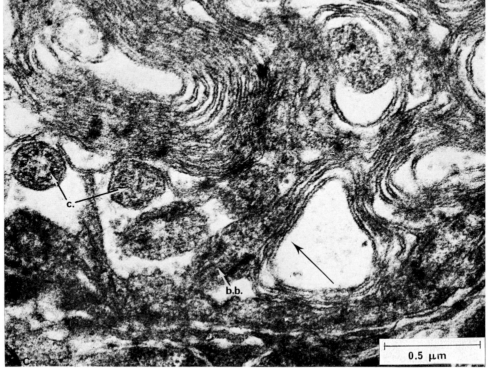

K. M. LYONS

Plate 3

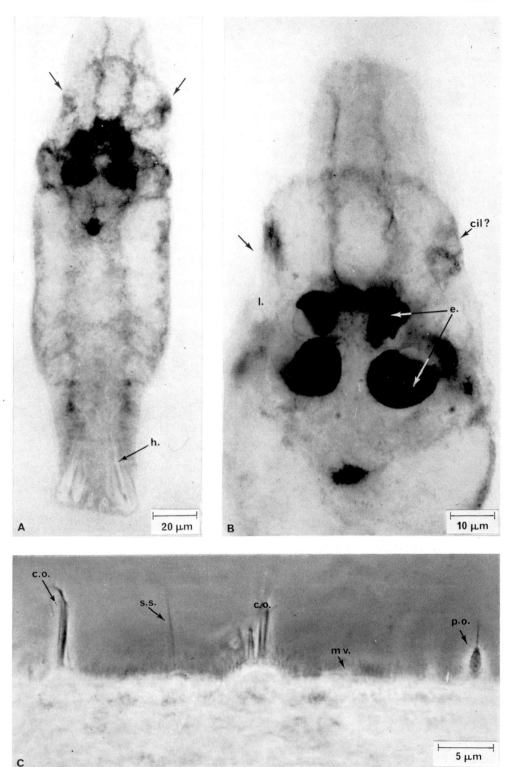

K. M. LYONS

PLATE 3

A. Photomicrograph of the oncomiracidium of *Entobdella soleae* showing the brain and nervous system which has been stained using the indoxyl acetate method for non-specific esterases. The arrows indicate stained areas which may represent the positions of the ciliary organs. Ventral view.

B. Photomicrograph showing the head region of *Entobdella soleae* oncomiracidium after staining with the indoxyl acetate method. Dorsal view showing the four pigmented eyespots each equipped with a lens. Arrows indicate indoxyl positive regions which may correspond in position to the ciliary presumed eyes located using the electron microscope.

C. Phase contrast photomicrograph of the lateral edge of the head of *Acanthocotyle lobianchi* fixed in 2.5% glutaraldehyde. Three different kinds of presumed sense organ can be seen; compound (multiciliate) sensilla, a peg organ ending in a single cilium and a single sensillum (tangoreceptor?). The epidermal microvilli found on the dorsal side of this worm are clearly visible.

Chemoreceptors in haematophagous insects

C. T. LEWIS

Department of Zoology and Applied Entomology,
Imperial College of Science and Technology, London

The distribution, structure and function of different types of sensilla concerned in olfaction and gustation in haematophagous insects are reviewed, with special reference to biting flies of the families Culicidae and Muscidae. In species of both families, receptors have been identified which are specialized for the perception of carbon dioxide and water vapour, stimuli common to all host animals. Discrimination between host species is most probably accomplished by the integration of information from other olfactory receptor cells which are excited differentially by a range of fatty acids and related compounds.

In female mosquitoes, different fields of contact chemoreceptors on the labella and labrum initiate nectar and blood feeding respectively, while the rates of ingestion of both are controlled by cibarial sensilla.

Functional aspects of the ultrastructure of olfactory receptor cells are briefly discussed with reference to receptor sites and accessory structures, the influence of dendrite branching on electrogenic and electrotonic events, and energy relationships within the cell.

CONTENTS

INTRODUCTION

Two main objectives can be distinguished in the study of chemoreception in haematophagous insects. In the first place, after identifying the different kinds of chemosensilla possessed by a species, we may determine the range of specific stimuli to which each is able to respond and thus obtain information which is essential to a full appreciation of host-seeking and feeding behaviour. Secondly, we may seek to understand the mechanisms within a sensory cell which enable it to carry out its function. Although much remains to be done, in both these fields of enquiry significant advances have been made in recent years, notably by the use of microelectrode and electron microscopic techniques.

The phenomenon of blood feeding is widely distributed in the Insecta, being found in the orders Anoplura, Thysanoptera, Hemiptera, Aphaniptera and Diptera. A few species of Lepidoptera are also blood feeders (review by Hocking, 1971). Most of the examples to be discussed in this paper will be taken from the Diptera; in particular from the families Culicidae and Muscidae in which the species habitually seek their hosts in free flight, in contrast to many insect blood feeders which live in close proximity to their hosts.

OLFACTORY SENSILLA

General characteristics

The complex of airborne stimuli emanating from warm-blooded vertebrate hosts, though variable in detail, conforms to a common general pattern dominated by carbon dioxide, water vapour, fatty acids and derivatives, ammonia, amines and thermal stimuli. The substances act as trans-specific chemical messengers, evoking reactions in the recipient species which are harmful to the emitter. They thus fall into the category of "kairomones", as defined by Brown, Eisner & Whittaker (1970), in contrast to allomones (trans-specific messenger substances evoking reactions favourable to the emitter) and pheromones (intraspecific messenger substances).

Olfactory sensilla are found on the antennae and in some species upon the palps. Each sensillum originates from a single epidermal cell which divides to form trichogen, tormagen, neurilemma and primary sense cells (Wigglesworth, 1953; Lawrence, 1966).

The cuticular structure produced by the trichogen cell is organized in one of two ways. The thin cuticle wall may be perforated by a dense array of submicroscopic pores (Plates 2A and 4A) a feature which is characteristic of trichoid, basiconic and club-shaped (clavate) sensilla. Alternatively, in coeloconic and styloconic sensilla the cuticle wall is relatively thick and provided with elongate clefts (Plate 1C, D) which in some species are apparently filled with a viscous material with staining properties which differentiate it from the cuticle wall proper (Lewis, 1970a).

There is no functional difference which can be consistently correlated with these two basic structural forms. In some species, e.g. *Apis mellifera* (Lacher, 1964), *Stomoxys calcitrans* (Lewis, in prep.), coeloconic or styloconic sensilla have been shown to respond to changes in concentration of carbon dioxide or water vapour, stimulants which differ from fugitive odours in being always present in the natural environment. But these structures are not unique in this respect, for some porous sensilla also respond to fluctuations in humidity and carbon dioxide concentration, for example clavate and basiconic sensilla of *Aedes aegypti* (Kellog, 1970) and *Stomoxys*; while the receptor cells of some coeloconic sensilla respond to a wide range of normally transient odours (e.g. in *Locusta migratoria,* Kafka, 1970).

Typically an olfactory sensillum possesses a number of bipolar sensory neurones, each behaving as a separate receptor unit having a characteristic response. The number present in a sensillum varies widely. For example, in

Stomoxys styloconic sensilla possess two receptor cells, trichoid sensilla, six or seven; in *Rhodnius prolixus* 15 may be associated with one sensillum (Wigglesworth & Gillett, 1934) (cf. Plate 4B). An extreme example is provided by the phytophagous bug *Pyrops candelaria* with about 200 sensory neurones to each plaque sensillum (Lewis & Marshall, 1970).

Receptor cells as information systems

Fundamentally the olfactory receptor cells of haematophagous insects are similar to those of other insects, the electrophysiological characteristics of which have been reviewed by Boeckh, Kaissling & Schneider (1965) and Schneider (1965, 1969). Microelectrode recordings from alongside single sensilla reveal both slow potential changes and the generation of impulses (Plate 1B). Many olfactory receptor cells discharge impulses regularly at a steady rate in the apparent absence of a stimulus. The slow receptor potential change recorded on stimulation by an odour may be positive or negative, suggesting a hyperpolarization or depolarization of the receptor membrane; this is correlated with a decreased or increased rate of generation of action potentials. Thus the activity of an olfactory cell may be increased, inhibited or unchanged by a given stimulus. For example, receptor cells of the sensilla basiconia have recently been identified on the palps of the Tsetse fly *Glossina austeni* which are excited by lower fatty acids and alcohols, inhibited by higher fatty acids and alcohols and unaffected by carbon dioxide (Lewis, unpubl.) (Plates 1B and 2B).

Certain olfactory receptor cells which exhibit a stereotyped response to a very narrow range of compounds have been termed "odour specialists" by Boeckh *et al.* (1965). These cells have the function of alerting an insect to "key odour stimuli" (Schneider, 1965) rather than that of contributing to the discrimination of many odours. The best documented examples are the sex pheromone receptors of male *Bombyx mori* silk moths, the cells of which can respond individually to single molecules of bombykol (Schneider, 1969). In blood sucking flies carbon dioxide receptors may be regarded as true odour specialists, though they have a much lower sensitivity than *Bombyx* pheromone receptors.

Other cells are considered to be odour generalists, each having a unique spectrum of responses to a wide range of odours, being excited by some, inhibited by some and unresponsive to others. Different generalist cells have different but partially overlapping reaction spectra. Thus, while a single cell cannot distinguish between different exciting stimuli, a group of such cells exhibits a large number of permutations and combinations of response pattern, acting together to provide a system capable of discriminating a large number of odours. The integration of such information in the deutocerebrum has yet to be studied, however.

The recognition by Schneider and his colleagues of the distinction between specialist and generalist receptor cells represents a major advance in the understanding of olfactory mechanisms. However, as more information becomes available, cells with responses intermediate between these extremes may be recognized. The fatty acid receptors of blood sucking flies are examples, which reveal some of the characteristics of both types and it is suggested that these may be termed "discriminating specialists" (see below).

Olfactory organs of Culicidae

The most nearly complete analysis of the performance of olfactory organs in a blood sucking insect has been achieved for the yellow fever mosquito, *Aedes aegypti*. A mosquito antenna consists of a scape, pedicellus and 13 flagellar segments. In a number of species, three morphological types of olfactory sensilla were recognized by Steward & Atwood (1963). These may be described as sharp trichoid (type A1), shorter, blunt trichoid (A2) and smaller, thorn-like basiconic sensilla (A3). Those of *Aedes aegypti* are illustrated in Plate 2D. In male mosquitoes only the terminal two segments possess such organs but in females they are present on all 13 flagellar segments; they are not uniformly distributed however, for types A1 and A3 increase in numbers distally, while type A2 sensilla are more numerous on the proximal segments. An account of the fine structure of two of these forms of sensilla in *Aedes* has been given by Slifer & Sekhon (1962). In species of *Anopheles*, coeloconic sensilla have also been found (Steward & Atwood, 1963).

In addition, the maxillary palpi are endowed with clavate sensilla (Roth & Willis, 1952; McIver & Charlton, 1970). Those of *Aedes aegypti* are illustrated in Plate 2C. The latter have been shown by Kellog (1970) in electrophysiological experiments to possess receptor cells which are very sensitive to carbon dioxide, giving a phasic response of 2-4 times the spontaneous impulse rate on exposure to an 0.01% change in concentration. Other receptor cells are also present in these clavate sensilla which respond to olfactory stimuli other than carbon dioxide. Kellog (1970) also showed that the A3 basiconic sensilla of *Aedes* antennae respond to water vapour, a change of 2% in relative humidity being detectable.

In olfactometer experiments, carbon dioxide has been shown to activate mosquitoes to fly while water vapour and warmth induce target orientation (Kellog & Wright, 1962) (see also Gillies, this volume p. 72). These three stimuli arise from all warm-blooded vertebrates; nevertheless species of mosquito show specific host preferences in settling behaviour; indeed, one individual out of a group of the same host species may be preferred. What is the sensory basis of this refined host discrimination?

"Discriminating specialist" cells

The evidence available strongly suggests that it lies with the receptors which respond to fatty acids and their derivatives. Lacher (1967) has shown that in *Aedes aegypti*, cells of the A1 trichoid sensilla are generally excited by fatty acids, while their resting activity is depressed by essential oils; those of A2 trichoid sensilla are excited by higher fatty acids, depressed by lower fatty acids and variable in response to essential oils.

Lacher suggests that the A1 receptors, being generally consistent in their responses, should be regarded as odour specialists. But an additional conclusion of some importance can be extracted from his results. If there were no inter-receptor differences in response pattern to different fatty acids, then different mixes of fatty acids arising from different hosts would not be readily distinguishable from each other—the intensity of response would vary only in the same way as it would for different concentrations of the same mixture. But the comparative results for six A1 sensilla given in Lacher (1967) reveal that there are significant differences in the reaction spectra of different sensilla towards a range of fatty acids, differences which are sufficient to provide a distinctive pattern of responses to different combinations of fatty acids. Thus these receptor cells behave to some extent as "specialist" cells, responding to a narrow range of similar compounds, but also within that range act as "generalist" cells providing a pattern of information which could be integrated to allow refined discrimination of fatty acid odours. In other words the results suggest they behave as "discriminating specialists".

The A2 cells are considered by Lacher (1967) to behave as odour generalists towards essential oils and as specialists towards fatty acids; here again it seems that the cells sensitive to fatty acids could more accurately be described as "discriminating specialists". It is the variation in response pattern to mixtures of fatty acids which on the present evidence provides the most probable sensory basis for detailed host discrimination.

Among the fatty acids, lactic acid may play a particularly significant role, for this has been shown by Acree, Turner, Gouck, Beroza & Smith (1968) and Muller (1968) to be active as an attractant in behavioural experiments with female *Aedes aegypti*; and the former authors observed a correlation between the relative attractiveness of three different humans and the quantities of *l*-lactic acid present in acetone washes of their hands.

Olfactory organs of higher Diptera

Compared with the multi-segmented antennae of the Culicidae and other Nematocera, the antennae of cyclorraphous Diptera are very compact structures bearing a complex, dense array of chemosensilla upon the enlarged third segments (e.g. Plate 1A). For example, the haematophagous stable fly *Stomoxys calcitrans* possesses approximately 5000 chemosensilla on the external surface of each antenna; among these are found two types of trichoid, three of basiconic, and one each of clavate and styloconic sensilla (Lewis, 1971). In addition there are two olfactory pits containing scores of structurally more delicate forms of basiconic and styloconic sensilla (cf. Plate 2A).

Electrophysiological experiments have gone some way towards unravelling the complexity of the information provided by these organs. The styloconic sensilla possess carbon dioxide specialist receptor cells; other cells responding to this stimulant are found in some of the basiconic and clavate sensilla, which also possess cells responding to humidity changes. Other receptor cells of basiconic and trichoid sensilla are sensitive to fatty acids, esters, alcohols, ammonia and amines (Lewis, in prep.). Although not strictly within the scope of this paper it is of interest that trichoid sensilla located at the distal extremity of the antenna also possess receptor cells which are highly sensitive to temperature changes, though not to infra-red radiation.

The antennae of Tsetse flies are similar in form to those of *Stomoxys* but possess smaller numbers of sensilla, fewer morphologically different types and more complex olfactory pits (Lewis, 1970b). Unlike *Stomoxys* however,

Glossina species possess numbers of fine basiconic sensilla on the outer surface of the maxillary palps (Lewis, unpubl.). These sensilla differ from those of the palps of *Aedes* mosquitoes in being unresponsive to carbon dioxide. This stimulant is known to be extremely important in the host-seeking behaviour of mosquitoes (v.s.) and *Stomoxys* (Gatehouse, 1970) but it is not yet clear whether it is an important stimulus for *Glossina* species (cf. Gatehouse, this volume, p. 93). The palpal organs of *Glossina* are, however, very sensitive to lower fatty acids. In addition to their role in host seeking and discrimination, these sensilla may contribute to the initiation of probing, for Hughes (1954) has shown that odours of butyric and valeric acid can induce probing in hungry tsetse flies.

CONTACT CHEMORECEPTORS

Contact chemoreceptors in mosquitoes are found grouped upon the tarsi, labella (Plate 3C), labrum and within the cibarium. Stimulation of any one of these receptor fields with an appropriate solution will bring about aspiration of fluid in hungry females (Owen, 1963; Salama, 1966). There are, however, significant differences between these groups in their sensitivity to different stimulants, as both these authors have pointed out. By stimulating different mouthparts separately with solutions offered in capillary tubes Salama (1966) was able to show that in *Aedes aegypti,* sensilla associated with the labella and cibarium respond to sugar solutions and water, whilst in females those of the labrum and the cibarium respond to blood. Thus nectar and blood feeding are initiated by separate contact chemoreceptor systems. For both fluids, however the rate of ingestion is controlled by the cibarial sensilla. In addition to the above, Hudson (1970) has observed structures resembling sensilla on the hypopharynx of three species. Their function is at present unknown.

The factor in blood which stimulates feeding in Tsetse flies was observed by Yorke and Blacklock as early as 1915 to be associated with blood cells, not plasma. Hosoi (1958, 1959) also observed that washed blood cells possessed the gorging stimulus for *Culex pipiens* and identified the active principle as adenosine-5-phosphate. ATP has since been shown to induce gorging in many other haematophagous insects, including *Rhodnius* (Friend, 1965), *Aedes* (Galun, Avidor & Bar-Zeev, 1963), *Tabanidae* (Lall, 1969), *Glossina* (Galun & Margalit, 1969).

In blood-feeding Muscidae, contact chemosensilla are also found upon the tarsi (Lewis, 1954) and on the outer surface of the labellum; other sensilla with much thicker walls are normally hidden among the armature of rasping teeth and are brought forward when the proboscis is everted to pierce the host skin (Plate 3A, B). In addition four pharyngeal chemosensilla are present (Rice, 1970). The fine structures of the outer labellar and tarsal sensilla of *Stomoxys calcitrans* have been described by Adams, Holbert & Forgash (1965) who observed that the dendrites of four or five receptor cells extend to the base of the hollow seta in an electron-dense sheath. The dendrite of one cell apparently ends at the base of the seta and is presumed to be a mechanoreceptor. The others extend to the tip of the hollow seta within a thick sheath, which fuses at one side to the wall of the seta. There are two fine pores through the cuticle at the tip of the sensillum (Lewis, 1970a)

In neither Culicidae nor biting Muscidae has the physiology of these organs been studied in detail by electrophysiological techniques. It is probable that, as in blowfly contact chemoreceptors, different receptor cells within a sensillum respond to different categories of stimuli; e.g. sugars, cations, water (review, Hodgson, 1968). Rees (1967) has put forward the interesting suggestion that the close fitting thick sheath around the dendrites may prevent the normal spread of receptor potential along the distal regions of the dendrite membrane; instead the outer lumen of the sensillum may provide a pathway for a form of saltatory conduction of potential from the receptor sites at the tip to the proximal dendrite region at the base of the hair, in which region action potentials are initiated.

FUNCTIONAL ASPECTS OF CHEMORECEPTOR ULTRASTRUCTURE

The established biophysical facts are that specific molecules received by the cell initiate the release of energy as a receptor potential change, with a correlated generation of action potentials. These events involve the movement of ions across the receptor cell membrane. To what extent can specific details of cellular fine structure be correlated with these processes?

Receptor sites

In most olfactory sensilla the cuticle is closely perforated by fine pores. There are minor variations in the organization of the terminal structures, but typically a number of tubules extend from the pores. These are cuticular in origin (Ernst, 1969) and of extremely fine dimensions, e.g. from 50 to 80Å internal diameter in *Stomoxys* (Lewis, 1971). They terminate in close apposition to the membranes of dendrite branches (Plate 4A). The interface between a tubule and dendrite is presumed to represent the location of a receptor site. In a typical *Stomoxys* basiconic sensillum there are about 20,000 such sites upon the dendrites of three receptor cells.

The tubules are filled with an aqueous medium probably secreted by the tormogen cell. Though small in volume (about 10^{-5} cubic micra per tubule in *Stomoxys*) the tubule medium plays an extremely important role in at least two processes. Firstly it must play an integral part in ion exchange activity across the membrane of the receptor site. Secondly, stimulating molecules must diffuse to the receptor site through the medium, which may thus possibly contribute to molecular discrimination. Though no details can be resolved in the electron microscope the tubule contents may be highly organized. Locke (1965) has suggested they may contain lipid/water liquid crystals.

The stimulating effect of an odorous molecule is transient and its removal from the active site must also be considered. Riddiford (1970) has shown that, after stimulation with tritiated pheromone molecules produced by females injected with tritiated acetate, a radioactive protein can be extracted from male *Antheraea* antennae. It is possible that the protein binding thus revealed may be part of the excitatory process, but in view of the stability of the bond it is also possible that it may be a mechanism for sequestering stimulant molecules from the active sites.

Dendrite organization

Elaborate patterns of subdivision of the outer dendrite region have been reported for olfactory receptor cells in *Stomoxys* and *Glossina* (Lewis 1970a, 1971). In *Stomoxys* the patterns range from the unbranched dendrites of trichoid sensilla through three patterns of multiple branching in basiconic sensilla to scroll-like concentric tight foldings of the dendrites in clavate sensilla. Although a full understanding of the significance of these complex variations in dendrite organization has yet to be achieved, at least one type of variation has been correlated with receptor sensitivity.

In many basiconic sensilla the dendrites are highly divided, forming a complex of more than 100 branches with a lumen less than 2 μm in diameter. What are the relative merits of a large unbranched dendrite compared with an array of smaller branches, as bioelectric components of a receptor system? Equations have been derived which relate fibre diameter to input impedance and decremental conduction of potential (e.g. review by Katz, 1966). From these it may be deduced that as the dendrite diameter is reduced by branching, for example by a factor of 10, the efficiency of passive electrotonic conduction of a receptor potential (the length constant) is reduced by a factor of approximately 3. But the input impedance rises by a factor of about 30; this means that the voltage change produced at a receptor site by a given stimulus will be appreciably greater in the smaller fibre and will more than compensate for greater conduction losses.

Thus a system of finely branched dendrites may be regarded as an adaptation to greater sensitivity, providing larger receptor potentials and, despite a less efficient electronic conduction, a greater potential change at the site of generation of impulses (Lewis, 1970a). This correlation of fine branching and sensitivity may well prove to be a phenomenon of very general application in sensory physiology.

The ciliary region

The dendrite of a chemoreceptor cell is clearly differentiated into two segments, the proximal or inner segment being appreciably greater in diameter than the outer segment. At the point of constriction between them a ciliary organelle is invariably found, with two basal bodies and rootlets extending into the inner segment and the nine paired peripheral fibrils extending into the outer segment (Plate 4B, C). A ciliary organelle is also found in the dendrites of insect mechanoreceptors and contact chemoreceptors, as well as in analogous external sensilla in other phyla. Its functional significance is uncertain but in olfactory cells one correlation which has been noted is with the microtubules which pass to the terminal regions of the outer dendrite (Plates 1D and 4C). These appear to arise by division from the ciliary fibrils (Slifer & Sekhon, 1969) and may have a skeletal function.

There is also some circumstantial evidence to suggest that the ciliary region may be the site of generation of action potentials; for the spreading receptor potential would be appreciably attenuated proximal to this region as the dendrite membrane suddenly expands in area and thus in capacity, in proportion to the square of the diameter (Lewis, 1970a). It is also evident that metabolic energy reserves are present in this region.

Energy relationships

The inner segments are abundantly supplied with mitochondria, up to the ciliary region. In comparison the outer dendrite segments possess no organelles other than microtubules; yet they must have access to a source of metabolic energy to compensate for the energy dissipated terminally in the generation of receptor potentials.

The small dimensions of the terminal dendrite branches indicate that the reserve of ions maintained there against a concentration gradient must be small and more immediately dependent upon metabolic recovery processes after electrogenic activity than the reserves of larger nerve fibres. The energy turnover in neurones, including the activity of the "sodium pump" (Caldwell & Keynes, 1957), involves compounds with high energy phosphate bonds produced by oxidative phosphorylation carried out in mitochondria. These considerations lead to the conclusion that the energy required to maintain the transduction process must be generated in the mitochondria of the inner segment and transported, probably as ATP, to the terminal dendrite branches; possibly via the microtubules.

The receptor cells are surrounded at all points by sheath cells, through which nutrients for the maintenance of the receptor must be admitted. The interrelationships of sheath cells and receptor cells are complex and not fully understood. The importance of their contribution to receptor site environment has already been mentioned. There may be two, three or four sheath cells surrounding a group of receptor cells in one sensillum. The inner cell may secrete a sheath or collar surrounding the dendrites below the level of the cuticle; apparently it also secretes the fluid medium surrounding the dendrites within the lumen, which by reason of its ionic composition must play a part in governing the excitability of the individual receptor cells.

ACKNOWLEDGEMENTS

It is a pleasure to acknowledge the skilled assistance of Mr R. Williams and Miss J. Finney in some of the work discussed in this review, which was aided by a research grant from the Overseas Development Administration.

I am also indebted to Dr T. A. M. Nash for supplies of *Glossina austeni* and Dr D. A. Griffiths for scanning electron microscope facilities.

REFERENCES

ACREE, F., TURNER, R. B., GOUCK, H. K., BEROZA, M. & SMITH, N., 1968. L-lactic acid: a mosquito attractant isolated from humans. *Science, N.Y., 161:* 1346-1347.

ADAMS, J. R., HOLBERT, P. E. & FORGASH, A. J., 1965. Electron microscopy of the contact chemoreceptors of the stable fly, *Stomoxys calcitrans* (Diptera, Muscidae). *Ann. ent. Soc. Am., 58:* 909-917.

BOECKH, J., KAISSLING, K. E. & SCHNEIDER, D., 1965. Insect olfactory receptors. *Cold Spring Harb. Symp. quant. Biol., 30:* 263-280.

BROWN, W. L., JR., EISNER, T. & WHITTAKER, R. H., 1970. Allomones and Kairomones: transpecific chemical messengers. *Bioscience, 20:* 21-22.

CALDWELL, P. C. & KEYNES, R. D. 1957. The utilisation of phosphate bond energy for sodium extrusion from giant axons. *J. Physiol., 137:* 12-13P.

ERNST, K-D., 1969. Die Feinstruktur von Reichsensillen auf der Antenne des Aaskafers *Necrophorus* (Coleoptera). *Z. Zellforsch., 94:* 72-102.

FRIEND, W. G., 1965. The gorging response in *Rhodnius prolixus* Ståhl. *Can. J. Zool., 43:* 125-132.

GALUN, R., AVI-DOR, Y. & BAR-ZEEV, M., 1963. Feeding response in *Aedes aegypti:* Stimulation by adenosine triphosphate. *Science, N.Y., 142:* 1674-1675.

GALUN, R. & MARGALIT, J., 1969. Adenosine nucleotides as feeding stimulants of the tsetse fly *Glossina austeni* Newst. *Nature, Lond., 222:* 583-584.

GATEHOUSE, A. G., 1970. Host finding by the stable fly, *Stomoxys calcitrans* (L.). *Proc. R. ent. Soc. Lond. (Ser C)., 34:* 37.

GATEHOUSE, A. G., 1972. In E. U. Canning & C. A. Wright (Eds), *Behavioural aspects of parasite transmission. Zool. J. Linn. Soc., 51,* Suppl. 1: 82-95.

GILLIES, M. T., 1972. In E. U. Canning & C. A. Wright (Eds), *Behavioural aspects of parasite transmission. Zool. J. Linn. Soc., 51,* Suppl. 1: 68-81.

HOCKING, B., 1971. Blood-sucking behaviour of terrestrial Arthropods. *A. Rev. Ent., 16:* 1-26.

HODGSON, E. S., 1968. Taste receptors of Arthropods. *Symp. zool. Soc. Lond., 23:* 269-277.

HOSOI, T., 1958. Adenosine-5'-phosphates as the stimulating agent in blood for inducing gorging in the mosquito. *Nature, Lond., 181:* 1664-1665.

HOSOI, T., 1959. Identification of blood components which induce gorging of the mosquito. *J. Insect Physiol., 3:* 191-218.

HUDSON, A., 1970. Notes on the piercing mouthparts of three species of mosquitoes viewed with the scanning electron microscope. *Can. Ent., 102:* 501-509.

HUGHES, J. C., 1957. Olfactory stimulation of tsetse flies and blowflies. *Bull. ent. Res., 48:* 561-579.

KAFKA, W. A., 1970. Molekulare Wechselwirkungen bei der Erregung einzeluer Riechzellen *Z. vergl. Physiol. 70:* 105-143.

KATZ, B., 1966. *Nerve, muscle and synapse..* New York: McGraw-Hill.

KELLOG, F. E., 1970. Water vapour and carbon dioxide receptors in *Aedes aegypti. J. Insect Physiol., 16:* 99-108.

KELLOG, F. E. & WRIGHT, R. H., 1962. The guidance of flying insects. V. Mosquito attraction. *Can. Ent., 94:* 1009-1016.

LACHER, V., 1964. Elektrophysiologische Untersuchungen an einzelnen rezeptoren fur Geruch, Kohlendioxyd, Luftfeuchtigkeit und Temperatur auf den antennen der Arbeitsbiene und der Drohne (*Apis mellifica* L.). *Z. vergl. Physiol., 48:* 587-623.

LACHER, V., 1967. Elecktrophysiologische Untersuchungen an einzelnen Geruchsrezeptoren auf den Antennen weiblicher Moskitos (*Aedes aegypti* L.). *J. Insect Physiol., 13:* 1461-1470.

LALL, S. B., 1969. Phagostimulants of haematophagous tabanids. *Entomologia exp. appl., 12:* 325-336.

LAWRENCE, P. A., 1966. Development and determination of hairs and bristles in the Milkweed Bug, *Onchopeltus fasciatus* (Lygaeidae, Hemiptera). *J. Cell Sci., 1:* 475-498.

LEWIS, C. T., 1954. Studies concerning the uptake of contact insecticides. I. The anatomy of the tarsi of certain Diptera of medical importance. *Bull. ent. Res., 45:* 711-722.

LEWIS, C. T., 1970a. Structure and function in some external receptors. *Symp. R. ent. Soc. Lond., 5:* 59-76.

LEWIS, C. T., 1970b. Preliminary observations on the fine structure of the olfactory receptors of Glossina species. *Proc. 1st. Symp. criaco Laboratorio da Tsetse e aplicacao pratica,* 299-303.

LEWIS, C. T., 1971. Superficial sense organs of the antennae of the fly, *Stomoxys calcitrans. J. Insect. Physiol., 17:* 449-461.

LEWIS, C. T. & MARSHALL, A. T., 1970. The ultrastructure of the sensory plaque organs of the Chinese lantern fly, *Pyrops candelaria* L. (Homoptera, Fulgoridae). *Tissue Cell, 2:* 375-385.

LOCKE, M., 1965. Permeability of insect cuticle to water and lipids. *Science, N.Y., 147:* 295-298.

McIVER, S. & CHARLTON, C., 1970. Studies on the sense organs on the palps of selected culicine mosquitoes. *Can. J. Zool., 48:* 293-295.

MULLER, W., 1968. Die Distanz—·.nd Kontakt—Orientierung der Stechmucken (*Aedes aegypti*) (Wirtsfindung, Stechverhalten und Blutemahlzeit). *Z. vergl. Physiol., 58:* 241-303.

OWEN, W. B., 1963. The contact chemoreceptor organs of the mosquito and their function in feeding behaviour. *J. Insect Physiol., 9:* 73-87.

REES, C. J. C., 1967. Transmission of receptor potential in Dipteran chemoreceptors. *Nature, Lond., 215:* 301-302.

RICE, M. J., 1970. Aspects of the structure, innervation and function of tsetse fly gut. *Trans. R. Soc. trop. Med. Hyg., 64:* 185.

RIDDIFORD, L. M., 1970. Antennal proteins of Saturniid moths—their possible role in olfaction. *J. Insect Physiol., 16:* 653-660.

ROTH, L. M. & WILLIS, E. R., 1952. Possible hygroreceptors in *Aedes aegypti* (L.) and *Blattella germanica* (L.) *J. Morphol., 91:* 1-14.

SALAMA, H. S., 1966. The function of mosquito taste receptors (*Aedes aegypti*). *J. Insect Physiol., 12:* 1051-1060.

SCHNEIDER, D., 1965. Chemical sense communication in insects. *Symp. Soc. exp. Biol., 20:* 273-297.

SCHNEIDER, D., 1969. Insect olfaction. Deciphering system for chemical messages. *Science, N.Y., 163:* 1031-1037.

SLIFER, E. H. & SEKHON, S. S., 1962. The fine structure of the sense organs on the antennal flagellum of the yellow fever mosquito *Aedes aegypti* (Linnaeus). *J. Morphol., 111:* 49-67.

SLIFER, E. H. & SEKHON, S. S., 1969. Some evidence for the continuity of ciliary fibrils and microtubules in the insect sensory dendrite. *J. Cell Sci., 4:* 527-540.

STEWARD, C. C. & ATWOOD, C. E., 1963. The sensory organs of the mosquito antenna. *Can. J. Zool., 41:* 577-594.

WIGGLESWORTH, V. B. & GILLETT, J. D., 1934. The function of the antennae in *Rhodnius prolixus* (Hemiptera) and the mechanism of orientation to the host. *J. exp. Biol., 11:* 120-139.

WIGGLESWORTH, V. B., 1953. The origin of sensory neurones in an insect, *Rhodnius prolixus* (Hemiptera). *Q. Jl miscrosc. Sci., 94:* 93-112.

YORKE, W. & BLACKLOCK, B., 1915. Food of *Glossina palpalis* in the Cape lighthouse peninsula, Sierra Leone. *Ann. trop. Med. Parasit., 9:* 363-380.

DISCUSSION

K. M. Lyons

Would Dr Lewis care to speculate about the function of the microtubules in sensory endings?

C. T. Lewis

Microtubules probably serve a number of different functions in different types of cell, but two speculations seem to be particularly relevant to transduction processes in sensory endings.

They may have a "skeletal" or mechanical function, being most evident in endings subject to deformation. Close arrays of microtubules are characteristic of insect mechanoreceptors. In tactile setae, for example, many microtubules form a closely knit structure (the "tubular organ") in the terminal region of the dendrite. I have suggested elsewhere (Lewis, 1970a) that this array may govern the viscous and elastic properties of the dendrite ending where it is stressed by the movement of the seta; i.e., in this special case the array of microtubules may play an important role in controlling the deformation of the transducer membrane and hence its sensitivity.

Secondly, microtubules may provide channels for the transport of ATP from the inner, metabolically active region of the dendrite to the peripheral regions, where energy is required for the maintenance of the dendrite in a state which will permit the generation of the receptor potential. In sensory endings, however finely branched the dendrites, at least one microtubule is always present in each branch.

S. M. Omer

Two general points I would like to pick up: the first, can you say hyperpolarization of the sensory neuron can mean, in behavioural terms, an accepted stimulus or not?; the second point is Schneider's classification of sensory neurons into "specialists" and "generalists" does not seem to fit when applied to host specificity of haematophagous insects, unlike responses to pheromones.

C. T. Lewis

If an olfactory sensory neurone is normally spontaneously active then a stimulus inducing hyperpolarization will, by inhibiting impulse generation, change the input to the central nervous system.

It is impossible to generalize on the resultant effect "in behavioural terms" for the motor responses of the insect will depend upon the integration of all afferent nervous activity; but in so far as the hyperpolarizing stimulus changes the input from certain categories of sensory neurone it will be producing effects which, when integrated with the input from other chemosensilla, could have positive results.

Referring to your second point: while carbon dioxide and water vapour receptor cells may be designated as true specialists, I agree that there is some ambiguity when the term is applied to receptor cells specialized for a *range* of related compounds, such as the carboxylic acids, esters and alcohols which appear to play an important role in the host specificity.

Here the qualified description "discriminating specialists" seems more appropriate, because such cells exhibit differential responses for specific compounds within the range, as discussed above (cf. Plate 1B).

D. J. Lewis

The sensilla of Phlebotomidae (sandflies) may be too small to study. Can one deduce that bunching of their palpal clavate sensilla, or elongation of antennal trichoid sensilla, indicate greater sensitivity?

C. T. Lewis

Larger numbers of specific olfactory sensilla and elongation of sensilla will both provide larger surface areas for the interception of odorous molecules and for this reason may be considered as adaptations favouring increased sensitivity. "Bunching" of receptors in more distal, exposed locations could also increase the efficiency of interception.

Of course other factors which govern sensitivity may also vary (e.g., density of pores in the receptor wall, inherent sensitivity of receptor cells, number of receptor cells) but among closely related species I think it would be reasonable to infer that the properties you describe favour increased sensitivity.

Plate 1

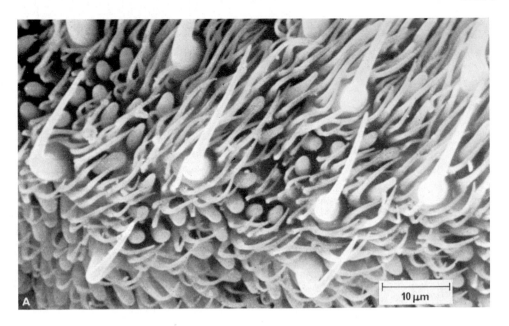

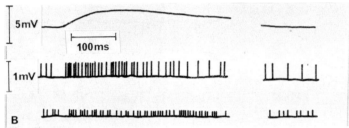

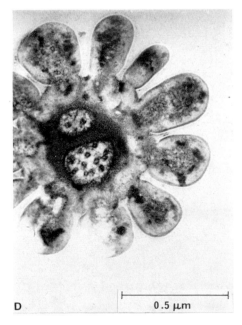

C. T. LEWIS

Plate 2

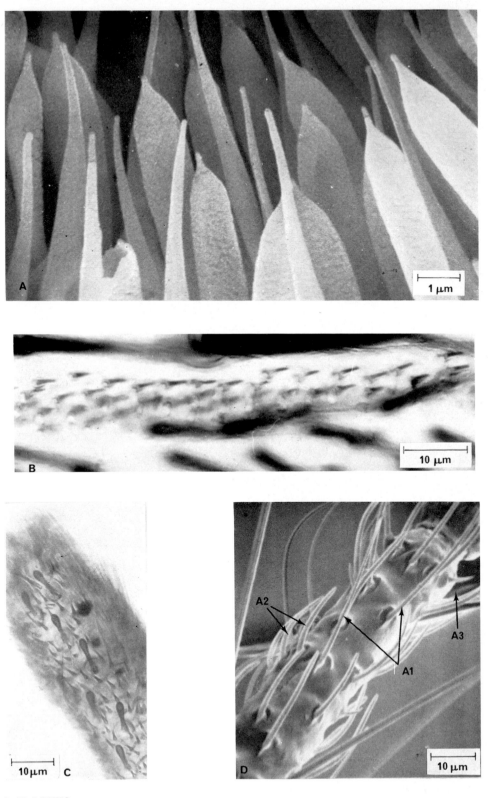

C. T. LEWIS

Plate 3

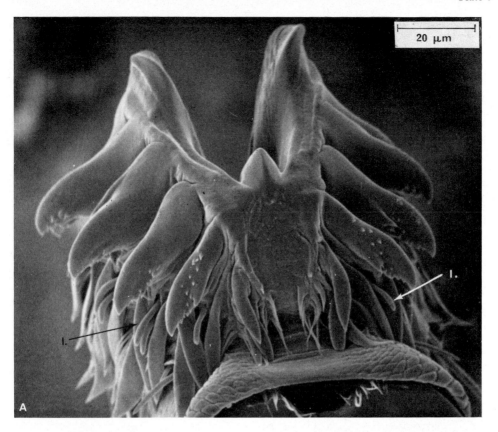

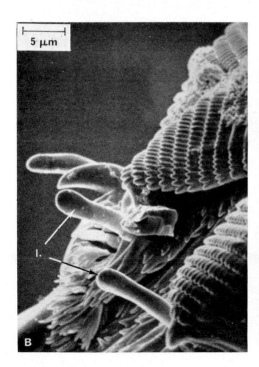

C. T. LEWIS

Plate 4

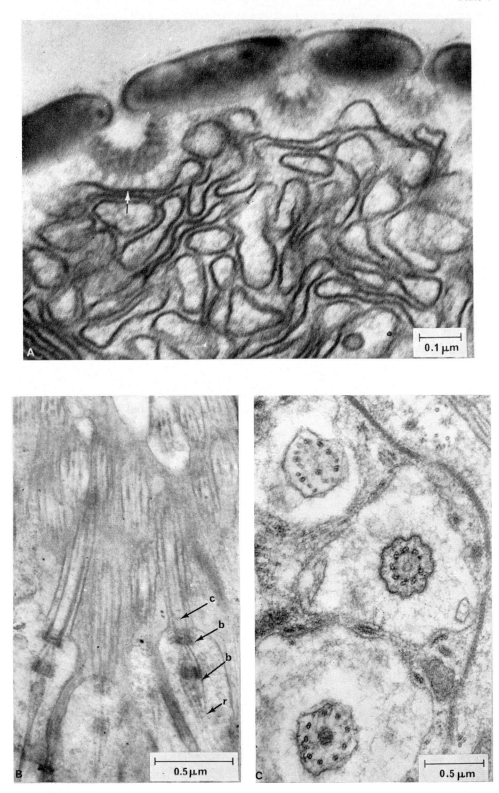

C. T. LEWIS

EXPLANATION OF PLATES

PLATE 1

A. Scanning electron micrograph of a ventral region of the third antennal segment, *Stomoxys calcitrans*. Trichoid sensilla and numerous basiconic sensilla interspersed with microtrichia.

B. Responses of a basiconic sensillum of the palps of the tsetse fly *Glossina austeni*, stimulated by a pulse of amyl alcohol odour. From an electrophysiological recording showing depolarizing compound receptor potential and coincident volleys of action potentials from two receptor cells having different adaptation rates.

C. Scanning electron micrograph of a styloconic sensillum on the antenna of *G. austeni*.

D. Electron micrograph of transverse section of styloconic sensillum, *G. austeni*. Stained with uranyl acetate and lead citrate.

PLATE 2

A. Stereoscan electron micrograph of a cluster of sensilla basiconica within a dissected olfactory pit of *G. austeni*. Fine perforations of the cuticular wall can be seen.

B. Light micrograph showing rows of fine sensilla basiconica along a palp of *G. austeni*, stained with cobalt sulphide.

C. Light micrograph of clavate sensilla on a palp of *Aedes aegypti*, stained with cobalt sulphide.

D. Scanning electron micrograph of part of antennal flagellum of ♀ *Aedes aegypti*. A1, A2, A3: different forms of olfactory sensilla (see text).

PLATE 3

A. Scanning electron micrograph of an everted labellum of *Stomoxys calcitrans*, showing the inner labellar contact chemoreceptors (l) among the armature of teeth.

B. Scanning electron micrograph of part of an everted labellum of *Glossina austeni*, showing contact chemoreceptors (l) adjacent to the rasping teeth.

C. Scanning electron micrograph showing contact chemoreceptors (l) on the labellum of the mosquito *Anophles stephensi*.

PLATE 4

A. Electron micrograph of an oblique section of a sensillum basiconicum of *S. calcitrans*, showing pores in the cuticle and fine tubules which appear to make contact with dendrites (arrow). Lead citrate. From Lewis (1971).

B. Electron micrograph of a longitudinal-oblique section through a group of receptor cells in an antennal sensillum of the bug *Rhodnius prolixus*, showing ciliary organelles between the inner and outer segments of dendrites. Uranyl acetate, lead citrate. b, Basal plates; c, ciliary fibrils; r, rootlets.

C. Electron micrograph of a transverse section in the ciliary region of receptor cells of a trichoid sensillum, *S. calcitrans*. Uranyl acetate, lead citrate.

Index

215